Lecture Notes in Physics

Springer
Berlin
Heidelberg
New York
Barcelona
Hong Kong
London
Milan
Paris
Singapore
Tokyo

Physics and Astronomy ONLINE LIBRARY

http://www.springer.de/phys/

The Editorial Policy for Proceedings

The series Lecture Notes in Physics reports new developments in physical research and teaching – quickly, informally, and at a high level. The proceedings to be considered for publication in this series should be limited to only a few areas of research, and these should be closely related to each other. The contributions should be of a high standard and should avoid lengthy redraftings of papers already published or about to be published elsewhere. As a whole, the proceedings should aim for a balanced presentation of the theme of the conference including a description of the techniques used and enough motivation for a broad readership. It should not be assumed that the published proceedings must reflect the conference in its entirety. (A listing or abstracts of papers presented at the meeting but not included in the proceedings could be added as an appendix.)

When applying for publication in the series Lecture Notes in Physics the volume's editor(s) should submit sufficient material to enable the series editors and their referees to make a fairly accurate evaluation (e.g. a complete list of speakers and titles of papers to be presented and abstracts). If, based on this information, the proceedings are (tentatively) accepted, the volume's editor(s), whose name(s) will appear on the title pages, should select the papers suitable for publication and have them refereed (as for a journal) when appropriate. As a rule discussions will not be accepted. The series editors and Springer-Verlag will normally not interfere with the detailed editing except in fairly obvious cases or on technical matters.

Final acceptance is expressed by the series editor in charge, in consultation with Springer-Verlag only after receiving the complete manuscript. It might help to send a copy of the authors' manuscripts in advance to the editor in charge to discuss possible revisions with him. As a general rule, the series editor will confirm his tentative acceptance if the final manuscript corresponds to the original concept discussed, if the quality of the contribution meets the requirements of the series, and if the final size of the manuscript does not greatly exceed the number of pages originally agreed upon. The manuscript should be forwarded to Springer-Verlag shortly after the meeting. In cases of extreme delay (more than six months after the conference) the series editors will check once more the timeliness of the papers. Therefore, the volume's editor(s) should establish strict deadlines, or collect the articles during the conference and have them revised on the spot. If a delay is unavoidable, one should encourage the authors to update their contributions if appropriate. The editors of proceedings are strongly advised to inform contributors about these points at an early stage.

The final manuscript should contain a table of contents and an informative introduction accessible also to readers not particularly familiar with the topic of the conference. The contributions should be in English. The volume's editor(s) should check the contributions for the correct use of language. At Springer-Verlag only the prefaces will be checked by a copy-editor for language and style. Grave linguistic or technical shortcomings may lead to the rejection of contributions by the series editors. A conference report should not exceed a total of 500 pages. Keeping the size within this bound should be achieved by a stricter selection of articles and not by imposing an upper limit to the length of the individual papers. Editors receive jointly 30 complimentary copies of their book. They are entitled to purchase further copies of their book at a reduced rate. As a rule no reprints of individual contributions can be supplied. No royalty is paid on Lecture Notes in Physics volumes. Commitment to publish is made by letter of interest rather than by signing a formal contract. Springer-Verlag secures the copyright for each volume.

The Production Process

The books are hardbound, and the publisher will select quality paper appropriate to the needs of the author(s). Publication time is about ten weeks. More than twenty years of experience guarantee authors the best possible service. To reach the goal of rapid publication at a low price the technique of photographic reproduction from a camera-ready manuscript was chosen. This process shifts the main responsibility for the technical quality considerably from the publisher to the authors. We therefore urge all authors and editors of proceedings to observe very carefully the essentials for the preparation of camera-ready manuscripts, which we will supply on request. This applies especially to the quality of figures and halftones submitted for publication. In addition, it might be useful to look at some of the volumes already published. As a special service, we offer free of charge LaTeX and TeX macro packages to format the text according to Springer-Verlag's quality requirements. We strongly recommend that you make use of this offer, since the result will be a book of considerably improved technical quality. To avoid mistakes and time-consuming correspondence during the production period the conference editors should request special instructions from the publisher well before the beginning of the conference. Manuscripts not meeting the technical standard of the series will have to be returned for improvement.

For further information please contact Springer-Verlag, Physics Editorial Department II, Tiergartenstrasse 17, D-69121 Heidelberg, Germany

Series homepage – http://www.springer.de/phys/books/lnpp

Tobias Brandes   (Ed.)

Low-Dimensional Systems

Interactions and Transport Properties

Lectures of a Workshop Held in Hamburg,
Germany, July 27-28, 1999

 Springer

SEP/ME
PHYS

Editor

Tobias Brandes
1. Institut für Theoretische Physik
Universität Hamburg
Jungiusstr. 9
20355 Hamburg, Germany

Library of Congress Cataloging-in-Publication Data applied for.

Die Deutsche Bibliothek - CIP-Einheitsaufnahme

Low dimensional systems : interactions and transport properties ;
lectures of a workshop held in Hamburg, Germany, July 27 - 28, 1999 /
Tobias Brandes (ed.). - Berlin ; Heidelberg ; New York ; Barcelona ;
Hong Kong ; London ; Milan ; Paris ; Singapore ; Tokyo : Springer,
2000
  (Lecture notes in physics ; 544)
  ISBN 3-540-67237-0

ISSN 0075-8450
ISBN 3-540-67237-0 Springer-Verlag Berlin Heidelberg New York

Springer-Verlag is a company in the BertelsmannSpringer publishing group.
© Springer-Verlag Berlin Heidelberg 2000
Printed in Germany

The use of general descriptive names, registered names, trademarks, etc. in this publication
does not imply, even in the absence of a specific statement, that such names are exempt
from the relevant protective laws and regulations and therefore free for general use.

Typesetting: Camera-ready by the authors/editor
Cover design: *design & production*, Heidelberg

Printed on acid-free paper
SPIN: 10720741   55/3144/du - 5 4 3 2 1 0

Preface

Experimental progress over the past few years has made it possible to test a number of fundamental physical concepts related to the motion of electrons in low dimensions. The production and experimental control of novel structures with typical sizes in the sub-micrometer regime has now become possible. In particular, semiconductors are widely used in order to confine the motion of electrons in two-dimensional heterostructures. The quantum Hall effect was one of the first highlights of the new physics that is revealed by this confinement. In a further step of the technological development in semiconductor-heterostructures, other artificial devices such as quasi one-dimensional 'quantum wires' and 'quantum dots' (artificial atoms) have also been produced. These structures again differ very markedly from three- and two-dimensional systems, especially in relation to the transport of electrons and the interaction with light. Although the technological advances and the experimental skills connected with these new structures are progressing extremely fast, our theoretical understanding of the physical effects (such as the quantum Hall effect) is still at a very rudimentary level.

In low-dimensional structures, the interaction of electrons with one another and with other degrees of freedoms such as lattice vibrations or light gives rise to new phenomena that are very different from those familiar in the bulk material. The theoretical formulation of the electronic transport properties of small devices may be considered well-established, provided interaction processes are neglected. On the other hand, the influence of interactions on quantities such as the conductance and conductivity remains one of the most controversial issues of recent years. Progress has been achieved partly in the understanding of new quasiparticles such as skyrmions, composite fermions, and new states of the interacting electron gas (e.g., Tomonaga–Luttinger liquids), both theoretically and in experiments. At the same time, it has now become clear that for fast processes in small structures not only the interaction but also the non-equilibrium aspect of quantum transport is of fundamental importance. It is also apparent now that, in order to understand a major part of the experimental results, transport theories are required that comprise both the non-equilibrium and the interaction aspect, formulated in the framework of a physical language that was born almost exactly one century ago: quantum mechanics.

This volume contains the proceedings of the 219th WEH workshop 'Interactions and transport properties of low dimensional systems' that took place on July 27 and 28, 1999, at the Warburg–Haus in Hamburg, Germany. Talks were given by leading experts who presented and discussed recent advances for the benefit of participants from all over the world, among whom were many young students. This is one reason why the present volume is more than simply a state-of-the-art collection of review articles on electronic properties of interacting lower dimensional systems. We have also tried to achieve a style of presentation that allows an advanced student or newcomer to use this as a textbook. Further study is facilitated by the many references at the end of each article. Thus we encourage all those interested to use this book together with pencil and sometimes the further reading, to gain an entry into this fascinating field of modern physics.

The articles in Part I present the physics of interacting electrons in one-dimensional systems. Here, one of the key issues is the identification of power-laws appearing as a function of energy scales such as the voltage, the frequency, or the temperature. A generic theoretical description of the physics of such systems is provided by the Tomonaga–Luttinger model, where in general power-law exponents depend on the strength of the electron–electron interaction. Further important issues are the proper definition of the conductance of interacting systems, the experimental verification of the predictions, and the search for new phases in quantum wires, as discussed in detail in the individual contributions.

The articles in Part II present an introduction to non-equilibrium transport through quantum dots, a survey of spin-related effects appearing in electronic transport properties, and new phenomena in two-dimensional systems under quantum Hall conditions, i.e. in strong magnetic fields.

All the contributions contain new and surprising results. One can definitely predict that many more novel aspects of the physics of 'interactions plus non-equilibrium in low dimensions' will emerge in the future. At this point, let me express the wish that this book will help to motivate readers to take part in this fascinating, rapidly developing field of physics. I would like also to use the present opportunity to thank all the participants and the speakers of the workshop for their contributions, and to acknowledge the friendly support of the WE Heraeus foundation.

Hamburg, November 1999   *Tobias Brandes*

Contents

Part I

Transport and Interactions in One Dimension

Nonequilibrium Mesoscopic Conductors Driven by Reservoirs

Akira Shimizu and Hiroaki Kato

Department of Basic Science, University of Tokyo, Komaba, Tokyo 153-8902, Japan

Abstract. In order to specify a nonequilibrium steady state of a quantum wire (QWR), one must connect reservoirs to it. Since reservoirs should be large 2d or 3d systems, the total system is a large and inhomogeneous 2d or 3d system, in which *e-e* interactions have the same strength in all regions. However, most theories of interacting electrons in QWR considered simplified 1d models, in which reservoirs are absent or replaced with noninteracting 1d leads. We first discuss fundamental problems of such theories in view of nonequilibrium statistical mechanics. We then present formulations which are free from such difficulties, and discuss what is going on in mesoscopic systems in nonequilibrium steady state. In particular, we point out important roles of energy corrections and non-mechanical forces, which are induced by a finite current.

1   Introduction

According to nonequilibrium thermodynamics, one can specify nonequilibrium states of macroscopic systems by specifying local values of thermodynamical quantities, such as the local density and the local temperature, because of the local equilibrium [1,2]. When one studies transport properties of a mesoscopic conductor (quantum wire (QWR)), however, the local equilibrium is not realized in it, because it is too small. Hence, in order to specify its nonequilibrium state uniquely, one must connect *reservoirs* to it, and specify their chemical potentials (μ_L, μ_R) instead of specifying the local quantities of the conductor (Fig. 1). The reservoirs should be large (macroscopic) 2d or 3d systems. Therefore, to really understand transport properties, we must analyze such a composite system of the QWR and the 2d or 3d reservoirs, Although the QWR itself may be a homogeneous 1d system, the total system is a *2d or 3d inhomogeneous system* without the translational symmetry. Moreover, *many-body interactions* are important *both* in the conductor and in the reservoirs: If electrons were free in a reservoir, electrons could neither be injected (absorbed) into (from) the conductor, nor could they relax to achieve the local equilibrium. However, most theories considered simplified 1d models, in which reservoirs are absent or replaced with noninteracting 1d leads [3–12].

In this paper, we study transport properties of a composite system of a QWR plus reservoirs, where *e-e* interactions are present in all regions. By critically reviewing theories of the conductance, we first point out fundamental problems of the theories in view of nonequilibrium statistical mechanics. We then present formulations which are free from such difficulties, and discuss what is going on in mesoscopic systems in nonequilibrium steady state. In particular, we point

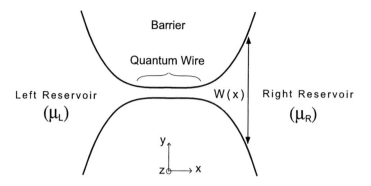

Fig. 1. A two-terminal conductor composed of a QWR and reservoirs.

out important roles of energy corrections and non-mechanical forces, which are induced by a finite current.

2 A Critical Review of Theories of the DC Conductance

In this section, we critically review theories of the DC conductance G of interacting electrons in a QWR. Note that two theories which predict different nonequilibrium states can (be adjusted to) give the same value of G (to agree with experiment). Hence, the comparison of the values of G among different theories is not sufficient. For definiteness, we consider a two-terminal conductor composed of a quantum wire (QWR) and two reservoirs (Fig. 1), which are defined by a confining potential u^c, at zero temperature. Throughout this paper, *we assume that u^c is smooth and slowly-varying, so that electrons are not reflected by u^c* (i.e., the wavefunction evolves adiabatically). We also assume that only the lowest subband of the QWR is occupied by electrons. A finite current I is induced by applying a finite difference $\Delta\mu = \mu_L - \mu_R$ of chemical potentials between the two reservoirs, and the DC conductance is defined by $G \equiv \langle I \rangle/(\Delta\mu/e)$ [13], where $\langle I \rangle$ is the average value of I.

Let us consider a clean QWR, which has no impurities or defects. For *noninteracting* electrons the Landauer-Büttiker formula gives $G = e^2/\pi\hbar$ [14], whereas G for *interacting* electrons has been a subject of controversy [15]. Most theories before 1995 [3–6] predicted that G should be "renormalized" by the *e-e* interactions as $G = K_\rho e^2/\pi\hbar$, where K_ρ is a parameter characterizing the Tomonaga-Luttinger liquid (TLL) [16–19]. However, Tarucha et al. found experimentally that $G \simeq e^2/\pi\hbar$ for a QWR of $K_\rho \simeq 0.7$ [20]. Then, several theoretical papers have been published to explain the absence of the renormalization of G [8–12,21]. Although they concluded the same result, $G = e^2/\pi\hbar$, the theoretical frameworks and the physics are very different from each other. Since most theories are based either on the Kubo formula [22] (or, similar ones based on the adiabatic switching of an "external" field), or on the scattering theory, we review these two types of theories critically in this section.

2.1 Problems and Limitations of the Kubo Formula when Applied to Mesoscopic Conductors

When one considers a physical system, it always interacts with other systems, R_1, R_2, $\cdots$, which are called heat baths or reservoirs. Nonequilibrium properties of the system can be calculated if one knows the reduced density matrix $\hat{\zeta} \equiv \mathrm{Tr}_{R1+R2+\cdots}[\hat{\zeta}_{\mathrm{total}}]$. Here, $\hat{\zeta}_{\mathrm{total}}$ is the density operator of the total system, and $\mathrm{Tr}_{R1+R2+\cdots}$ denotes the trace operation over reservoirs' degrees of freedom. To find $\hat{\zeta}$, Kubo [22] assumed that the system is initially in its equilibrium state. Then an "external field" $\mathbf{E}_{\mathrm{ext}}$ is applied adiabatically (i.e., $\mathbf{E}_{\mathrm{ext}} \propto e^{-\epsilon|t|}$), *which is a fictitious field* because it does not always have its physical correspondence (see below). The time evolution of $\hat{\zeta}$ was calculated using the von Neumann equation of an isolated system; i.e., *it was assumed that the system were isolated from the reservoirs* during the time evolution [2]. Because of these two assumptions (the fictitious field and isolated system), some conditions are required to get correct results by the Kubo formula. To examine the conditions, we must distinguish between *non-dissipative responses* (such as the DC magnetic susceptibility) and *dissipative responses* (such as the DC conductivity σ). The non-dissipative responses are essentially equilibrium properties of the system; in fact, they can be calculated from *equilibrium* statistical mechanics.

For non-dissipative responses, Kubo [22,23] and Suzuki [24] established the conditions for the validity of the Kubo formula, by comparing the formula with the results of equilibrium statistical mechanics: (i) The proper order should be taken in the limiting procedures of $\omega, q \to 0$ and $V \to \infty$, where ω and q are the frequency and wavenumber of the external field, and V denotes the system volume. (ii) The dynamics of the system should have the following property;

$$\lim_{t \to \infty} \langle \hat{A}\hat{B}(t) \rangle_{\mathrm{eq}} = \langle \hat{A} \rangle_{\mathrm{eq}} \langle \hat{B} \rangle_{\mathrm{eq}}, \tag{1}$$

where $\langle \cdots \rangle_{\mathrm{eq}}$ denotes the expectation value in the thermal equilibrium, and $\hat{A}$ and $\hat{B}$ are the operators whose correlation is evaluated in the Kubo formula. Any integrable models do not have this property [24,26–28]. Hence, *the Kubo formula is not applicable to integrable models, such as the Luttinger model*, even for (the simple case of) non-dissipative responses [24].

For dissipative responses, the conditions for the applicability of the Kubo formula would be stronger. Unfortunately, however, they are not completely clarified, and we here list some of known or suggested conditions for σ:
(i′) Like as condition (i), the proper order should be taken in the limiting procedures. For σ the order should be [25]

$$\sigma = \lim_{\omega \to 0} \lim_{q \to 0} \lim_{V \to \infty} \sigma_{\mathrm{formula}}(q, \omega; V). \tag{2}$$

(ii′) Concerning condition (ii), a stronger condition seems necessary for dissipative responses: The closed system that is taken in the calculation of the Kubo formula should have the thermodynamical stability, i.e., it approaches the thermal equilibrium when it is initially subject to a macroscopic perturbation. (Otherwise, it would be unlikely for the system to approach the correct steady state

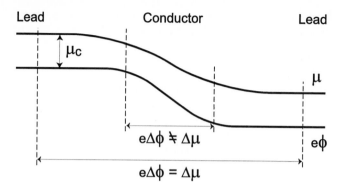

(a) macroscopic inhomegeneous conductor

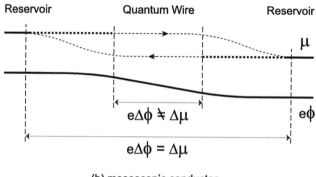

(b) mesoscopic conductor

Fig. 2. Schematic plots of the chemical potential μ [30] and the electrostatic potential ϕ, for (a) a macroscopic inhomogeneous conductor and (b) a mesoscopic conductor. For case (a), the local equilibrium is established, and thus μ and ϕ can be defined in all regions. The differences $e\Delta\phi$ and $\Delta\mu$ are not equal if one takes the differences between both ends of the conductor, whereas $e\Delta\phi = \Delta\mu$ if the differences are taken between the leads. For case (b), μ cannot be defined in the QWR and boundary regions (although in some cases μ could be defined separately for left- and right-going electrons), whereas ϕ can be defined in all regions. Similarly to case (a), $e\Delta\phi \neq \Delta\mu$ if one takes the differences between both ends of the QWR, whereas $e\Delta\phi = \Delta\mu$ if the differences are taken between the reservoirs.

in the presence of an external field.) In classical Hamiltonian systems, this condition is almost equivalent to the "mixing property" [26–28], which states that Eq. (1) should hold for *any* $\hat{A}$ and $\hat{B}$, where $\langle\cdots\rangle_{\mathrm{eq}}$ is now taken as the average over the equi-energy surface. It is this condition, rather than the "ergodicity", that guarantees the thermodynamical stability [26–28]. Although real physical systems should always have this property, some theoretical models do not. In particular, any integrable models do not have this property [26–28].

(iii') We here suggest that all driving forces, including non-mechanical ones, should be identified [29]. In fact, the formula gives the current density in the following form,

$$\langle \mathbf{J} \rangle = \sigma_{\text{formula}} \mathbf{E}_{\text{ext}}, \tag{3}$$

whereas the *exact* definition of σ is given by nonequilibrium thermodynamics as [1,2]

$$\langle \mathbf{J} \rangle = -\sigma \nabla (\mu/e) - L_{12} \nabla \beta = \sigma \mathbf{E} - \sigma \nabla (\mu_{\text{c}}/e) - L_{12} \nabla \beta. \tag{4}$$

Here, β denotes the inverse temperature, μ is the "chemical potential" which consists of a chemical portion μ_{c} and the electrostatic potential ϕ [1,30];

$$\mu = \mu_{\text{c}} + e\phi \quad \text{(hence, } \Delta\mu = \Delta\mu_{\text{c}} + e\Delta\phi \text{ for differences)}. \tag{5}$$

Hence, to evaluate σ, one must find the relation between $\mathbf{E}_{\text{ext}}$ and $\mathbf{E}$, $\nabla\mu_{\text{c}}$ and $\nabla\beta$. In *homogeneous* systems, it is expected that $\nabla\mu_{\text{c}} = \nabla\beta = 0$, hence it is sufficient to find the relation between the fictitious field $\mathbf{E}_{\text{ext}}$ and the real field $\mathbf{E}$ [2,10,31]. In *inhomogeneous* systems, however, $\nabla\mu_{\text{c}} \neq 0$ and/or $\nabla\beta \neq 0$ in general [32], as shown in Fig. 2 (a). Therefore, one must find the relation between $\mathbf{E}_{\text{ext}}$ and these "non-mechanical forces" [29,33]. (See section 5.)

Unfortunately, these conditions are not satisfied in theories based on simplified models of mesoscopic systems. For example, the Luttinger model [17] used in much literature does not satisfy conditions (i) and (ii) because it is integrable. To get reasonable results, subtle procedures, which have not been justified yet, were taken in actual calculations. Moreover, the non-mechanical forces have not been examined, although they would be important because a mesoscopic conductor (a QWR plus reservoirs) is an inhomogeneous system.

We also mention limitation of the Kubo formula: it cannot be applied to the nonequilibrium noise (NEN), which is the current fluctuation in the presence of a finite current $\langle I \rangle$ $(= G\Delta\mu)$ [7,34–38]. The NEN at low frequency, $\langle \delta I^2 \rangle^{\omega \simeq 0}$, is usually proportional to $|\langle I \rangle| \propto |\Delta\mu|$. However, the Kubo formula assumes power series expansion about $\Delta\mu = 0$, hence cannot give any function of $|\Delta\mu|$ [29,39].

In sections 3 and 4, we present other formulations which are free from these problems and limitations. These formulations clarify what is going on in nonequilibrium mesoscopic conductors, because one can find the nonequilibrium steady state. This is impossible by the Kubo formula because it evaluates correlation functions in the *equilibrium* state.

2.2   Scattering-Theoretical Approaches

In view of many problems and limitations of the Kubo formula, it is natural to try to generalize Landauer's theory [14] to treat conductors with many-body interactions. Namely, the DC conductance may be given in terms of the scattering matrix (S matrix) for interacting electrons [7,11,34].

The advantages of the scattering-theoretical approaches may be as follows: (i) Neither the translation of $\Delta\phi_{ext}$ into $\Delta\mu$ nor the subtle limiting procedures of ω, q and V is necessary. (ii) There is no need for the mixing property of the 1d Hamiltonian $\hat{H}_1$. Hence, $\hat{H}_1$ can be the Hamiltonian of integrable 1d systems such as the TLL. (iii) In contrast to the Kubo formula, one can calculate the NEN [7,34–36].

However, to define the S matrix, one must define incoming and outgoing states. Although they can be defined trivially for free electrons, it is nontrivial in the presence of many-body interactions. In high-energy physics, they are defined based on the *asymptotic condition*, which assumes that particles behave like free (but renormalized) ones as $t \to \pm\infty$, i.e., before and after the collision [40]. For example, an electron (in the vacuum) before or after the collision becomes a localized "cloud" of electrons and positrons, which extend only over the Compton length, and this cloud can be regarded as a renormalized electron. In condensed-matter physics, on the other hand, the asymptotic condition is not satisfied for electrons in metals and doped semiconductors. In fact, elementary excitations (Landau's quasi particles) are accompanied with the backflow, which extends *all over the crystal* [41], in contradiction to the asymptotic condition. Because of this fundamental difficulty, the scattering approaches to mesoscopic conductors replaced the reservoirs with 1d leads in which electrons are free [7,11,34]. Therefore, real reservoirs, in which electrons behave as 2d or 3d interacting electrons, have not been treated by the scattering-theoretical approaches.

3   Combined Use of Microscopic Theory and Thermodynamics [21]

The basic idea of this method is as follows: Since a QWR is a small system, and is most important, it should be treated with a full quantum theory. On the other hand, reservoirs are large systems whose dynamics is complicated, hence it could be treated with thermodynamics (in a wide sense). Utilizing these observations, we shall develop thermodynamical arguments to find the nonequilibrium steady state that is realized when a finite $\Delta\mu$ is applied between the reservoirs. This is the key of this method because when the steady state is found, G (and other observables) can be calculated by straightforward calculations. Although in some cases *formal* calculations can be performed without finding the steady state [12], we stress that such formal theories are incomplete because another theory is required to relate $\Delta\mu$ of such theories with $\Delta\mu$ of the reservoirs, by which G is defined.

An advantage of the present method is that we do not need to find the relation between $\Delta\phi_{ext}$ and $\Delta\mu$ because $\langle I \rangle$ is directly calculated as a function of $\Delta\mu$. Another advantage is that it is applicable to NEN and nonlinear responses because nonequilibrium steady state is directly obtained.

3.1   Conductance of the 1d Fermi Liquid

It is generally believed that a 1d interacting electron system is not the Fermi liquid (FL) [41], but the Tomonaga-Luttinger liquid (TLL) [16–19]. For this reason, many papers on 1d systems [3–6,8,9,11,12] use the word FL to indicate *non-interacting* electrons, i.e., a Fermi *gas*. However, we do not use such a misleading terminology; by a FL we mean *interacting* quasi-particles. Since the backflow is induced by the interaction [41], the Landauer's argument of non-interacting particles [14] cannot be applied to a FL. On the other hand, real systems have finite length and finite intersubband energies, in contradiction to the assumptions of the TLL. Hence, some real systems might be well described as a FL. Therefore, G of a FL is non-trivial and interesting [15]. Furthermore, we will show in section 5 that the results for the FL suggest very important phenomena that is characteristic to nonequilibrium states of inhomogeneous systems. Note also that the following calculations look similar to the derivation of fundamental relations in the theory of the FL [41]. However, G of mesoscopic conductors was not calculated in such calculations. The most important point to evaluate G is to find the nonequilibrium steady state.

We find the nonequilibrium steady state using a thermodynamical argument as follows: In the reservoirs, electrons behave as a 2d or 3d (depending on the thickness of the reservoir regions) FL. Since we have assumed that u^c is smooth and slowly-varying, a 2d or 3d quasi-particle in a reservoir, *together with its backflow*, can evolve adiabatically into a 1d quasi-particle and its backflow in the QWR, without reflection. In this adiabatic evolution, the quasi-particle mass m^* and the Landau parameters f also evolve adiabatically, and the energy is conserved. Therefore, quasi-particles with $\varepsilon(k > 0) \leq \mu_{\mathrm{L}}$ are injected from the left reservoir. Here, ε is the quasi-particle energy;

$$\varepsilon(k) = \frac{\hbar^2 k^2}{2m^*} + \frac{\hbar}{\mathcal{L}} \sum_{k'} f(k, k') \delta n(k'), \qquad (6)$$

where $\delta n(k) \equiv n(k) - \Theta(|k| \leq k_{\mathrm{F}})$, with $n(k)$ being the quasi-particle distribution. The last term of this expression represents energy correction by interactions among quasi-particles [41]. On the other hand, a quasi-hole below μ_{L} should not be injected because otherwise the recombination of a quasi-particle with the quasi-hole would produce excess entropy, in contradiction with the principle of minimum entropy production. Similarly, quasi-particles with $\varepsilon(k < 0) \leq \mu_{\mathrm{R}}$ are injected from the right reservoir, with no quasi-holes are injected below μ_{R}. Therefore, the nonequilibrium steady state under a finite $\Delta\mu = \mu_{\mathrm{L}} - \mu_{\mathrm{R}}$ should be the "shifted Fermi state", in which quasi-particle states with $\varepsilon(k \geq 0) \leq \mu_{\mathrm{L}}$ and $\varepsilon(k < 0) \leq \mu_{\mathrm{R}}$ are all occupied. Hence, the right- (left-) going quasi-particles have the chemical potential $\mu_+ = \mu_{\mathrm{L}}$ ($\mu_+ = \mu_{\mathrm{R}}$). Considering also the charge neutrality, we can write the distribution function as

$$n(k) = \Theta(|k - q| \leq k_{\mathrm{F}}). \qquad (7)$$

Then, Eq. (6) yields

$$\mu_\pm = \frac{\hbar^2 k_F^2}{2m^*} \pm \hbar q \left[\frac{\hbar k_F}{m^*} + \frac{f_{++} - f_{+-}}{2\pi} \right],$$ (8)

where $f_{++} \equiv f(k_F, k_F)$ and $f_{+-} \equiv f(k_F, -k_F) = f(-k_F, k_F)$. Hence,

$$\Delta\mu = 2\hbar q \left[\frac{\hbar k_F}{m^*} + \frac{f_{++} - f_{+-}}{2\pi} \right].$$ (9)

On the other hand, considering the spin degeneracy, $\langle I \rangle$ is calculated as

$$\langle I \rangle = 2e\frac{q}{\pi} \left[\frac{\hbar k_F}{m^*} + \frac{f_{++} - f_{+-}}{2\pi} \right].$$ (10)

Here, the $f_{+\pm}$-dependent terms represent the backflow. Since the same factor appears in Eq. (9), we find that the conductance is independent of m^* and $f_{+\pm}$;

$$G \equiv \frac{\langle I \rangle}{\Delta\mu/e} = \frac{e^2}{\pi\hbar}.$$ (11)

Since we have identified the nonequilibrium steady state, we can calculate not only G but also other nonequilibrium properties such as the NEN [21].

It is instructive to represent Eqs. (8)-(10) in terms of the bare parameters. As in the case of 3d Fermi liquid [41], we can show that [42]

$$\frac{\hbar k_F}{m} = \frac{\hbar k_F}{m^*} + \frac{f_{++} - f_{+-}}{2\pi}.$$ (12)

Hence, we can rewrite Eq. (10) as $\langle I \rangle = 2e(q/\pi)(\hbar k_F/m)$. Therefore, quasi particles (whose group velocity is $\hbar k_F/m^*$) plus their backflows carry exactly the same current as the bare particles, for the same q, i.e., for the same shifted Fermi distribution. On the other hand, Eq. (8) is rewritten as $\mu_\pm = \hbar^2 k_F^2/2m^* \pm \hbar^2 q k_F/m$. Although $\mu_\pm \neq [\mu_\pm$ of bare particles], $\Delta\mu = [\Delta\mu$ of bare particles] for the same q. These facts result in the independence of G on the Landau parameters.

3.2 Conductance of the Tomonaga–Luttinger Liquid [21]

We now consider a clean TLL. The low-energy dynamics of a TLL is described by the charge (ρ) and spin (σ) excitations (whose quantum numbers are N_q^ρ and N_q^σ, respectively, where $q \neq 0$ denotes the wavenumber), and the zero modes (quantum numbers $N_\pm^\rho$, $N_\pm^\sigma$) [16–19]. The eigenenergy is given by

$$E = \sum_{\nu=\rho,\sigma} v^\nu \sum_q \hbar|q|N_q^\nu + \frac{\pi\hbar}{2\mathcal{L}} \sum_{\nu=\rho,\sigma} [v_N^\nu(N_+^\nu + N_-^\nu)^2 + v_J^\nu(N_+^\nu - N_-^\nu)^2],$$ (13)

where $v_N^\nu = v^\nu/K_\nu$ and $v_J^\nu = K_\nu v^\nu$ ($\nu = \rho, \sigma$). Here, the parameters v^ν and K_ν are renormalized by the e-e interactions (except that $K_\sigma = 1$ by the SU(2) symmetry). The DC current is given by

$$\langle I \rangle = 2ev_J^\rho(N_+^\rho - N_-^\rho)/\mathcal{L}.$$ (14)

We apply a thermodynamical argument to find the nonequilibrium steady state. Unlike the FL case, there is no adiabatic continuity between the TLL in the QWR and the FL in the reservoirs. We therefore argue differently: In the linear response regime the steady state must be the state with the minimum energy among states which satisfy given external conditions. Otherwise, the system would be unstable and would evolve into a state with lower energy. For our purpose, it is convenient to take the value of $\langle I \rangle$ as the given external condition. Then, from Eqs. (13) and (14), we find that the steady state should be the state with $N_q^\rho = N_q^\sigma = 0$ (for all q), $N_+^\rho + N_-^\rho = 0$, $N_+^\sigma + N_-^\sigma = N_+^\sigma - N_-^\sigma = 0$, and $N_+^\rho - N_-^\rho > 0$. This state may be called the "shifted Fermi state" of the TLL. Furthermore, in the steady state, electrons in the left reservoir and right-going electrons in the TLL should be in the "chemical equilibrium", in which electrons in the FL phase are transformed into right-going electrons in the TLL phase at a constant rate. Therefore, their chemical potentials should be equal [43];

$$\mu_{\mathrm{L,R}} = \mu_{+,-}^\rho \equiv \frac{\partial E}{\partial N_{+,-}^\rho} = \frac{\pi\hbar}{\mathcal{L}}[v_N^\rho(N_+^\rho + N_-^\rho) \pm v_J^\rho(N_+^\rho - N_-^\rho)], \qquad (15)$$

where we have used Eq. (13). Hence,

$$\Delta\mu = \mu_{\mathrm{L}} - \mu_{\mathrm{R}} = \frac{2\pi\hbar}{\mathcal{L}}v_J^\rho(N_+^\rho - N_-^\rho). \qquad (16)$$

By dividing Eq. (14) by this expression, we obtain the same result for G as Eq. (11), in agreement with experiment [20].

Since we have identified the nonequilibrium steady state, we can calculate not only G but also other nonequilibrium properties such as the NEN [21].

4   Projection Theory [44,45]

Although we have successfully found the nonequilibrium steady state of interacting electrons in section 3, a possible objection against the formulation may be that the theory is rather intuitive. In this section, we present a full statistical-mechanical theory, which is free from such an objection. In this theory, we start from the Hamiltonian of 3d interacting electrons confined in the composite system of the QWR and reservoirs, Fig. 1. This original system is projected onto an effective 1d system, and the equation of motion for the reduced density operator of the 1d system is derived. From this equation, we can find the nonequilibrium steady state as a function of $\Delta\mu$ between the reservoirs. This allows us to evaluate various nonequilibrium properties.

4.1   Decomposition of the 3d Electron Field [44]

We start from the 3d electron field $\hat{\psi}(\mathbf{r})$ subject to a confining potential $u^c(\mathbf{r})$ (which defines the QWR and two reservoirs connected to it), impurity potential

$u^i(\mathbf{r})$ (whose average $\overline{u^i}$ is absorbed in $u^c(\mathbf{r})$, hence $\overline{u^i} = 0$), external electrostatic potential $\phi_{\text{ext}}(\mathbf{r})$, and the e-e interaction of equal strength $v(\mathbf{r}-\mathbf{r}')$ in all regions:

$$\hat{H} = \int d^3r \, \hat{\psi}^\dagger(\mathbf{r}) \left[-\frac{\hbar^2}{2m}\nabla^2 + u^c(\mathbf{r}) + u^i(\mathbf{r}) + e\phi_{\text{ext}}(\mathbf{r}) \right] \hat{\psi}(\mathbf{r})$$
$$+ \frac{1}{2} \int d^3r \int d^3r' \, \hat{\rho}(\mathbf{r})v(\mathbf{r}-\mathbf{r}')\hat{\rho}(\mathbf{r}'), \tag{17}$$

where $\hat{\rho}(\mathbf{r})$ is the charge density. We will find the nonequilibrium steady state for $\Delta\mu > 0$. For this state, $\langle\hat{\rho}(\mathbf{r})\rangle \neq 0$, which gives rise to a long-range force. We extract it as the renormalization of the electrostatic potential

$$e\phi(\mathbf{r}) = e\phi_{\text{ext}}(\mathbf{r}) + \int d^3r' \, v(\mathbf{r}-\mathbf{r}')\langle\hat{\rho}(\mathbf{r}')\rangle, \tag{18}$$

and a c-number V_{av}. Namely, $\hat{H}$ is recast in terms of $\delta\hat{\rho}(\mathbf{r}) \equiv \hat{\rho}(\mathbf{r}) - \langle\hat{\rho}(\mathbf{r})\rangle$ as

$$\hat{H} = \int d^3r \, \hat{\psi}^\dagger(\mathbf{r}) \left[-\frac{\hbar^2}{2m}\nabla^2 + u^c(\mathbf{r}) + u^i(\mathbf{r}) + e\phi(\mathbf{r}) \right] \hat{\psi}(\mathbf{r})$$
$$+ \frac{1}{2} \int d^3r \int d^3r' \, \delta\hat{\rho}(\mathbf{r})v(\mathbf{r}-\mathbf{r}')\delta\hat{\rho}(\mathbf{r}') + V_{\text{av}}. \tag{19}$$

To decompose $\hat{\psi}(\mathbf{r})$, we consider the single-body part of $\hat{H}$. Recall that $u^c(\mathbf{r})$ is assumed to be smooth and slowly-varying, to avoid undesirable reflections at the QWR-reservoir boundaries. In this case, the single-body Schrödinger equation

$$\left[-\frac{\hbar^2}{2m}\nabla^2 + u^c(\mathbf{r}) + u^i(\mathbf{r}) + e\phi(\mathbf{r}) \right] \varphi(\mathbf{r}) = \varepsilon\varphi(\mathbf{r}) \tag{20}$$

has solutions that propagate through the QWR having the energy $\varepsilon \simeq \varepsilon_{\text{F}}$ [46];

$$\varphi_k(\mathbf{r}) \simeq \frac{1}{\sqrt{\mathcal{L}}} \exp\left[i \int_0^x K_k(x)dx \right] \varphi^\perp(y, z; x). \tag{21}$$

Here, $\varphi^\perp(y, z; x)$ is the wavefunction of the lowest subband at x, representing the confinement in the lateral (yz) directions, and $\mathcal{L}$ is the normalization length in the x direction. All the other modes are denoted by $\varphi_\nu(\mathbf{r})$, which includes solutions that are localized in either reservoir, and extended solutions whose ε are not close to ε_{F}. Since any function of $\mathbf{r}$ can be expanded in terms of $\varphi_k(\mathbf{r})$'s and $\varphi_\nu(\mathbf{r})$'s, so is the $\mathbf{r}$ dependence of the electron field operator;

$$\hat{\psi}(\mathbf{r}) = \sum_k \hat{c}_k \varphi_k(\mathbf{r}) + \sum_\nu \hat{d}_\nu \varphi_\nu(\mathbf{r}) \equiv \varphi^\perp(y, z; x)\hat{\psi}_1(x) + \hat{\psi}_{\text{R}}(\mathbf{r}). \tag{22}$$

The 3d electron field has thus been decomposed into the 1d field and the 3d field $\hat{\psi}_{\text{R}}(\mathbf{r})$, which we call the "reservoir field." It can be decomposed into the low-energy components $\hat{\psi}_{\text{RL}}$ and $\hat{\psi}_{\text{RR}}$ (which are localized in the left and right reservoirs, R_{L} and R_{R}, respectively) and the high-energy component $\hat{\psi}_{\text{RH}}$ as

$$\hat{\psi}_{\text{R}} = \hat{\psi}_{\text{RL}} + \hat{\psi}_{\text{RR}} + \hat{\psi}_{\text{RH}}. \tag{23}$$

For low-energy phenomena, we can take $\hat{\psi}_{\text{R}} = \hat{\psi}_{\text{RL}} + \hat{\psi}_{\text{RR}}$.

4.2 Hamiltonian for the 1d and the Reservoir Fields [44]

By expressing $\hat{H}$ in terms of $\hat{\psi}_1$ and $\hat{\psi}_R$, we obtain the single-body part of $\hat{\psi}_1$ (denoted by $\hat{H}_1^0$), the $\hat{\psi}_1$-$\hat{\psi}_1$ interaction ($\hat{V}_{11}^0$), the $\hat{\psi}_1$-$\hat{\psi}_R$ interaction ($\hat{V}_{1R}^0$), the $\hat{\psi}_R$-$\hat{\psi}_R$ interaction ($\hat{V}_{RR}^0$), and the single-body part of $\hat{\psi}_R$ ($\hat{H}_R^0$). By the screening effect of $\hat{V}_{RR}^0$, $\hat{V}_{1R}^0$ is renormalized as the screened interaction $\hat{V}_{1R}$. Similarly, by the screening effect of $\hat{V}_{1R}$, $\hat{V}_{11}^0$ is renormalized as the screened interaction $\hat{V}_{11}$. We therefore recast $\hat{V}_{11}^0 + \hat{V}_{1R}^0 + \hat{V}_{RR}^0$ as $\hat{V}_{11} + \hat{V}_{1R} + \hat{V}_{RR}$, where $\hat{V}_{1R}$ and $\hat{V}_{RR}$ have no screening effects on $\hat{V}_{11}$ and $\hat{V}_{1R}$. In this way, $\hat{H}$ is decomposed as

$$\hat{H} = \hat{H}_1 + \hat{V}_{1R} + \hat{H}_R, \tag{24}$$

where $\hat{H}_1 \equiv \hat{H}_1^0 + \hat{V}_{11}$ is the Hamiltonian for $\hat{\psi}_1(x)$, $\hat{H}_R \equiv \hat{H}_R^0 + \hat{V}_{RR}$ is the one for $\hat{\psi}_R(\mathbf{r})$, and $\hat{V}_{1R}$ is the interaction between $\hat{\psi}_1(x)$ and $\hat{\psi}_R(\mathbf{r})$. In particular, $\hat{H}_1$ is evaluated as

$$\hat{H}_1 = \int dx\, \hat{\psi}_1^\dagger(x) \left[-\frac{\hbar^2}{2m} \frac{\partial^2}{\partial x^2} + u^\perp(x) + u_1^i(x) \right] \hat{\psi}_1(x)$$
$$+ \frac{1}{2} \int dx \int dx'\, \delta\hat{\rho}_1(x,t) v_{11}(x,x') \delta\hat{\rho}_1(x',t). \tag{25}$$

Here, $u^\perp$ is the subband energy [44], $\delta\hat{\rho}_1(x,t) \equiv \hat{\psi}_1^\dagger(x)\hat{\psi}_1(x) - \langle \hat{\psi}_1^\dagger(x)\hat{\psi}_1(x) \rangle$ is the density fluctuation of the 1d field, and

$$u_1^i(x) \equiv \iint dydz\, |\varphi^\perp(y,z;x)|^2\, u^i(\mathbf{r}), \tag{26}$$

$$v_{11}(x,x') \equiv \iiiint dydzdy'dz'\, |\varphi^\perp(y,z;x)|^2 v^{sc}(\mathbf{r},\mathbf{r}')\, |\varphi^\perp(y',z';x')|^2 \tag{27}$$

are the impurity and two-body potentials for the 1d field, where v^{sc} is the screened two-body potential for $\hat{\psi}$.

It is seen that $u_1^i(x)$ is the average of the random potential $u_1^i(\mathbf{r})$ over the lateral wavefunction $\varphi^\perp$, which, as a function of y, is localized in a region of width $\sim W(x)$ for each x. Here, $W(x)$ denotes the width of the region in which electrons are confined (Fig. 1). From these observations, we can show that

$$u_1^i(x) \sim u^i(\mathbf{r}) \quad \text{for } x \in \text{QWR}, \tag{28}$$
$$|u_1^i(x)| \propto 1/\sqrt{W(x)} \quad \text{for } x \in \text{a reservoir}. \tag{29}$$

In a similar manner, the two-body potential for the 1d field behaves as

$$v_{11}(x,x') \lesssim v^{sc}([|x - x'|^2 + W(0)^2]^{1/2}) \quad \text{for } x \text{ or } x' \in \text{QWR}, \tag{30}$$
$$v_{11}(x,x') \sim (r^{sc}/W)\overline{v^{sc}}, \quad \text{for } x \sim x' \in \text{a reservoir}. \tag{31}$$

where r^{sc} denotes the range of v^{sc}, and $\overline{v^{sc}}$ the average of v^{sc} in the region $|\mathbf{r} - \mathbf{r}'| \lesssim r^{sc}$. We now assume that the width of the reservoirs is very large;

$$W(x) \to \infty \quad \text{as } x \to \pm\infty. \tag{32}$$

Then, Eqs. (29)-(31) yield $u_1^i(x) \to 0$ as $x \to \pm\infty$, and $v_{11}(x, x') \to 0$ as x or $x' \to \pm\infty$. Namely, $\hat{H}_1$ represents interacting $\hat{\psi}_1(x)$ field that gets free as $x \to \pm\infty$. On the other hand, the interaction $\hat{V}_{1R}$ between $\hat{\psi}_1(x)$ and $\hat{\psi}_R(\mathbf{r})$ becomes stronger in the reservoir regions, whereas it is negligible in the QWR because at low energies $\hat{\psi}_R(\mathbf{r})$ does not penetrate into the QWR. Therefore, the 1d field $\hat{\psi}_1(x)$ is subject to different scatterings in different regions of x: In the QWR, $\hat{\psi}_1$ is scattered by the $\hat{\psi}_1$-$\hat{\psi}_1$ interaction and the impurity potentials, whereas $\hat{\psi}_1$ is excited and attenuated by the reservoir field in the reservoir regions through the $\hat{\psi}_1$-$\hat{\psi}_R$ interactions (Fig. 3).

4.3 Equation of Motion for the Reduced Density Operator [44,45]

We have successfully rewritten the Hamiltonian $\hat{H}$ in terms of $\hat{\psi}_1$ and $\hat{\psi}_R$. We must go one step further because $\hat{H} = \hat{H}_1 + \hat{V}_{1R} + \hat{H}_R$ describes very complicated dynamics, and thus the von Neumann equation for the density operator $\hat{\zeta}$,

$$i\hbar\frac{\partial}{\partial t}\hat{\zeta}(t) = \left[\hat{H}_1 + \hat{V}_{1R} + \hat{H}_R, \hat{\zeta}(t)\right] \tag{33}$$

is impossible to solve. *This unsolvability guarantees the thermodynamical stability (mixing property) of the total system* [26–28]. We turn this fact to our own advantage, and reduce the theory to a tractable one. The basic idea is as follows: $\hat{H}_1$ describes the 1d correlated electrons, and is most important. Hence, it should be given a full quantum-mechanical treatment. Concerning $\hat{V}_{1R}$, on the other hand, multiple interactions by $\hat{V}_{1R}$ seem unimportant. Hence, we may treat it by a second-order perturbation theory [47]. For $\hat{H}_R$, it describes the 2d or 3d interacting electrons, for which many properties are well known, and we can utilize the established results. Moreover, since the reservoirs are large, we can assume the *local equilibrium*: both reservoirs are in their equilibrium states with the chemical potentials μ_L and μ_R, respectively. We denote the reduced density operator of the reservoir field for this local equilibrium state by $\hat{\zeta}_R$.

From these observations, we may project out the reservoir field $\hat{\psi}_R$ as follows. Consider the reduced density operator for the 1d field; $\hat{\zeta}_1(t) \equiv \mathrm{Tr}_R[\hat{\zeta}(t)]$. Up to

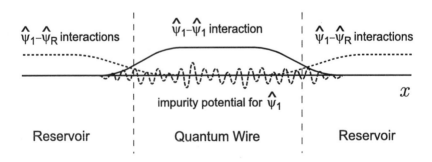

Fig. 3. Schematic diagram of the strengths of scatterings of the 1d field.

the second order in $\hat{V}_{1R}$ [47], the equation of motion of $\hat{\zeta}_1$ in the interaction picture of $\hat{H}_1 + \hat{H}_R$, is evaluated as

$$\frac{\partial}{\partial t}\hat{\zeta}_1(t) = \frac{-1}{\hbar^2}\int_{-\infty}^{t}dt'\mathrm{Tr_R}\left(\left[\hat{V}_{1R}(t),\left[\hat{V}_{1R}(t'),\hat{\zeta}_R\hat{\zeta}_1(t)\right]\right]\right), \tag{34}$$

where we have used the fact that $\hat{\zeta}_1(t)$ in the interaction picture varies only slowly, so that $\hat{\zeta}_1(t') \simeq \hat{\zeta}_1(t)$ in the correlation time of $\langle\hat{V}_{1R}(t)\hat{V}_{1R}(t')\rangle$ [48]. This equation represents that $\hat{\zeta}_1$ is driven by two reservoirs, which have different chemical potentials μ_L and μ_R, through the $\hat{\psi}_1$-$\hat{\psi}_R$ interaction. Since the trace is taken over the reservoir field, Eq. (34) is a closed equation for $\hat{\psi}_1$ and $\hat{\zeta}_1$. Its steady solution represents the nonequilibrium steady state of the 1d field driven by the reservoirs.

4.4   Current of the 1d Field [44]

We now turn to observables. We are most interested in the *total* current $\hat{I}$ which is given by $\hat{I}(x,t) \equiv \int\int dydz\hat{J}_x(\mathbf{r},t)$, where $\hat{J}_x$ denotes the x component of the current density. Note that $\hat{I}$ is different from the current of the 1d field defined by

$$\hat{I}_1(x,t) \equiv \frac{e}{2m}\left[\hat{\psi}_1^\dagger(x,t)\left\{\frac{\hbar}{i}\frac{\partial}{\partial x}\hat{\psi}_1(x,t)\right\} + \mathrm{h.c.}\right]. \tag{35}$$

The expectation values of $\hat{I}$, $\hat{I}_1$, and $\hat{I}_R$ (the current carried by the reservoir field) are schematically plotted in Fig. 4, which shows that $\langle I\rangle$ is mainly carried by $\langle I_1\rangle$ in the QWR and by $\langle I_R\rangle$ in the reservoirs, respectively. The transformation between $\hat{I}_1$ and $\hat{I}_R$ is caused by $\hat{V}_{1R}$, and thus $\hat{I}_1$ is *not* conserved: $\frac{\partial}{\partial x}\hat{I}_1 + \frac{\partial}{\partial t}\hat{\rho}_1 \neq 0$. At first sight, these facts might seem to cause difficulties in calculating $\langle I\rangle$ and $\langle\delta I^2\rangle$ from $\hat{\zeta}_1$. Fortunately, however, we can show that for any (nonequilibrium) steady state $\langle I\rangle$ and $\langle\delta I^2\rangle^{\omega\simeq0}$ are independent of x, and that

$$\langle I\rangle = \langle I_1\rangle \qquad\qquad \text{at } x \simeq 0 \tag{36}$$
$$\langle\delta I^2\rangle^{\omega\simeq0} = \langle\delta I_1^2\rangle^{\omega\simeq0} \qquad \text{at } x \simeq 0 \tag{37}$$

where $x = 0$ corresponds to the center of the QWR. Therefore, to calculate $\langle I\rangle$ and $\langle\delta I^2\rangle$, it is sufficient to calculate $\langle I_1\rangle$ and $\langle\delta I_1^2\rangle^{\omega\simeq0}$ at $x \simeq 0$, which can be calculated from $\hat{\zeta}_1$. Therefore, we have successfully reduced the 3d problem, Eq. (17), into the effective 1d problem, Eqs. (24), (34) and (35).

Actual calculations can be conveniently performed as follows. Although $\langle I\rangle$ and $\langle\delta I^2\rangle$ have both low- and high-frequency components, we are only interested in the low-frequency components, which are denoted by $\bar{I}(t)$ and $\delta\bar{I}^2(t)$, respectively. They are given by

$$\bar{I}(t) = \int_{-\infty}^{\infty}dt'f(t'-t)\mathrm{Tr}[\hat{I}(t')\hat{\zeta}(t')], \tag{38}$$

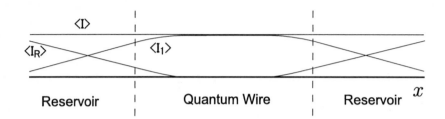

Fig. 4. Schematic plots of the expectation values $\langle I \rangle$, $\langle I_1 \rangle$, and $\langle I_R \rangle$, of the currents carried by $\hat{\psi}$, $\hat{\psi}_1$, and $\hat{\psi}_R$, respectively.

and similarly for $\delta \bar{I}^2(t)$. Here, $f(t'-t)$ is a filter function that is finite only in the region $t - \tau/2 \lesssim t' \lesssim t + \tau/2$, where $1/\tau \simeq$ the highest frequency of interest. From Eqs. (24), (34)-(38), we can construct the equations for $\bar{I}(t)$ and $\delta \bar{I}^2(t)$. They can be solved more easily than the equation for $\hat{\zeta}_1$, and the solutions fully describe the low-frequency behaviors of $\langle I \rangle$ and $\langle \delta I^2 \rangle$.

In the following, we present the results for $\bar{I}(t)$ for the case of impurity scatterings and for the the case of e-e interactions.

4.5   Application of the Projection Theory to the Case where Impurity Scatterings are Present in All Regions [45]

When electrons are scattered by impurities (one-body potentials) in all regions including reservoirs, whereas many-body scatterings are negligible, the $\hat{\psi}_1$-$\hat{\psi}_R$ interaction is given by

$$\hat{V}_{1R} = \int d^3r \hat{\psi}_1^\dagger(x)\varphi^{\perp*}(y,z;x)u^i(\mathbf{r})\hat{\psi}_R(\mathbf{r}) + \text{h.c.} \tag{39}$$

If we put $\hat{Y}_\alpha(\mathbf{r}) \equiv \varphi^{\perp*}(y,z;x)u^i(\mathbf{r})\hat{\psi}_{R_\alpha}(\mathbf{r})$ ($\alpha = $ L, R), then Eq. (34) becomes

$$\frac{\partial}{\partial t}\hat{\zeta}_1(t) = \frac{-1}{\hbar^2}\sum_\alpha \int_{-\infty}^t dt' \int d^3r \int d^3r' \left\{ \left\langle \hat{Y}_\alpha^\dagger(\mathbf{r},t)\hat{Y}_\alpha(\mathbf{r}',t') \right\rangle_\alpha \left[\hat{\psi}_1(x,t), \hat{\psi}_1^\dagger(x',t')\hat{\zeta}_1(t) \right] \right.$$

$$\left. + \left\langle \hat{Y}_\alpha(\mathbf{r},t)\hat{Y}_\alpha^\dagger(\mathbf{r}',t') \right\rangle_\alpha \left[\hat{\psi}_1^\dagger(x,t), \hat{\psi}_1(x',t')\hat{\zeta}_1(t) \right] \right\} + \text{h.c.} \tag{40}$$

where $\langle \cdots \rangle_\alpha$ denotes the expectation value for the equilibrium state of reservoir R_α, which has the chemical potential μ_α. After careful calculations using Eqs. (35)-(38), and considering the spin degeneracy, we find

$$\frac{d}{dt}\bar{I}(t) = -\gamma \left[\bar{I}(t) - \bar{I}_{\text{steady}} \right], \tag{41}$$

where $\bar{I}_{\text{steady}} \equiv (e/\pi\hbar)T\Delta\mu$, and $\gamma \simeq (2\pi/\hbar)n_{\text{imp}}|u^i|^2 \mathcal{D}_F$. Here, n_{imp} is the impurity density, u^i denotes the potential of an impurity ($u^i(\mathbf{r}) = \sum_\ell u^i\delta(\mathbf{r}-\mathbf{r}_\ell)$), and $\mathcal{D}_F$ is the density of states per unit volume, $\mathcal{D}(\mu_L) \simeq \mathcal{D}(\mu_R) \equiv \mathcal{D}_F$. It is seen that $\bar{I}$ approaches $\bar{I}_{\text{steady}}$ as $t \to \infty$. Therefore, the DC conductance is given by

$$G = T \frac{e^2}{\pi \hbar}, \tag{42}$$

in agreement with the Landauer-Büttiker formula [14]. Moreover, we find that *the steady state is stable*: For any (small) deviation from the steady state, $\bar{I}$ relaxes to the value $\bar{I}_{\text{steady}}$, with the relaxation constant γ.

4.6 Application of the Projection Theory to the Case where *e-e* Scatterings are Present in All Regions [45]

When the *e-e* interaction is important in all regions including reservoirs, whereas impurity scatterings are negligible, we find that the most relevant term of the $\hat{\psi}_1$-$\hat{\psi}_R$ interaction is given by

$$\hat{V}_{1R} = \int d^3r \int d^3r' \hat{\psi}_1^\dagger(x) \varphi^{\perp *}(y,z;x) \hat{\psi}_R(\mathbf{r}) v^{\text{sc}}(\mathbf{r},\mathbf{r}') \hat{\psi}_R^\dagger(\mathbf{r}') \hat{\psi}_R(\mathbf{r}') + \text{h.c.} \tag{43}$$

By this interaction, an electron is scattered into the QWR through the collision of two electrons in a reservoir, or, an electron in the QWR is absorbed in a reservoir. By putting $\hat{Y}_\alpha(\mathbf{r}) \equiv \int d^3r' \varphi^{\perp *}(y,z;x) \hat{\psi}_{R_\alpha}(\mathbf{r}) v^{\text{sc}}(\mathbf{r},\mathbf{r}') \hat{\psi}_{R_\alpha}^\dagger(\mathbf{r}') \hat{\psi}_{R_\alpha}(\mathbf{r}')$ (for $\alpha = $ L, R), which differs from $\hat{Y}_\alpha$ of the impurity-scattering case, we obtain the equation of motion for $\hat{\zeta}_1$ in the same form as Eq. (40). To derive the equation of motion for $\bar{I}$ from that equation, we need to calculate the correlation functions in the reservoirs, $\langle \hat{Y}_\alpha(t) \hat{Y}_\alpha^\dagger(t') \rangle_\alpha$ and $\langle \hat{Y}_\alpha^\dagger(t) \hat{Y}_\alpha(t') \rangle_\alpha$. We can easily calculate them using well-known results for the 2D or 3D FL because the reservoir electrons are believed to be the 2D or 3D FL. We also need correlation functions of the 1d field. They are quite different depending on the nature (FL or TLL) of the electrons in the QWR.

In the case where $\hat{\psi}_1$ behaves as a 1d FL, we obtain the equation for $\bar{I}(t)$ in the same form as Eq. (41), but now γ is a function of the *e-e* interaction parameters, and

$$\bar{I}_{\text{steady}} = 2 \frac{e}{\mathcal{L}} \sum_{k>0} \frac{\hbar k}{m} [\Theta(\mu_L - \varepsilon(k)) - \Theta(\mu_R - \varepsilon(-k))]. \tag{44}$$

Here, m is the *bare* mass, Θ is the step function, and $\varepsilon(\pm k)$ denotes the 1d quasi-particle energy, Eq. (6), *in the shifted Fermi state*. Note that if we simply took $\varepsilon(\pm k) = \hbar^2 k^2 / 2m^*$, then $\bar{I}_{\text{steady}} = (m^*/m)(e/\pi\hbar)\Delta\mu$, hence the conductance would be renormalized by the factor m^*/m. However, the correct expression (6) shows that $\varepsilon(\pm k)$ *are modified in the presence of a finite current*, and the correction terms are proportional to $q \propto \bar{I}$. As a result, the injection of an electron becomes easier or harder as compared with the case of $\bar{I} = 0$. This automatically "calibrates" the number of injected electrons, and we obtain

$$\bar{I}_{\text{steady}} = 2 \frac{\hbar k_F}{m} \left[\frac{\hbar k_F}{m^*} + \frac{1}{2\pi}(f_{++} - f_{+-}) \right]^{-1} \frac{e}{2\pi\hbar} \Delta\mu = \frac{e}{\pi\hbar} \Delta\mu, \tag{45}$$

where we have used Eq. (12). Therefore, $G = e^2/\pi\hbar$. Here, the interaction parameters of the 1d field are canceled in G, and those of the reservoir field are absorbed in γ. These observations confirm the results of section 5: the shifted Fermi state is realized as the nonequilibrium steady state, and the conductance is quantized.

The application of the projection theory to the case where $\hat\psi_1$ behaves as a TLL will be a subject of future study.

4.7   Advantages of the Projection Theory

A disadvantage of the projection theory is that calculations of G become rather hard as compared with the simple theories that are reviewed in section 2. However, the simple theories have many problems and limitations, as discussed there. The projection theory is free from such problems and limitations, and has the following advantages: (i) The value of $\langle I \rangle$ for the nonequilibrium state is directly calculated as a function of $\Delta\mu$. Hence, neither the translation of $\Delta\phi_{\text{ext}}$ into $\Delta\mu$ nor the subtle limiting procedures of ω, q and V is necessary. (ii) There is no need for the mixing property of the 1d Hamiltonian $\hat H_1$. Hence, $\hat H_1$ can be the Hamiltonian of integrable 1d systems such as the TLL. (iii) In contrast to the Kubo formula, which evaluates transport coefficients from equilibrium fluctuations, the projection theory gives the nonequilibrium steady state. This allows us to discuss what 1d state is realized and how the current is injected from the reservoirs. Moreover, we can calculate the NEN and nonlinear responses. (iv) The projection theory can describe the relaxation to the nonequilibrium steady state. This allows us to study the stability and the relaxation time of the nonequilibrium state.

5   Appearance of a Non-mechanical Force

We here discuss the applicability of the Kubo formula to inhomogeneous systems. The general conclusion of this section is independent of natures (such as a FL or TLL) and the dimensionality of the electron system. Hence, we will use the results for the 1d FL, which are obtained in section 3.1 and confirmed by a full statistical-mechanical theory in section 4.6.

The original form of the Kubo formula gives a conductivity that corresponds to the following conductance; $\langle I \rangle/\Delta\phi_{\text{ext}} \equiv G_{\text{Kubo}}$ [2,22]. Izuyama suggested that the conductance should be $\langle I \rangle/\Delta\phi \equiv G_{\text{Izuyama}}$, by considering the screening of ϕ_{ext} [10,31]. On the other hand, the exact definition of the conductance is $G \equiv \langle I \rangle/\Delta\mu$ [1,2]. For macroscopic inhomogeneous conductors, $e\Delta\phi \neq \Delta\mu$ in general if one takes the differences between both ends of the conductor, as sketched in Fig. 2(a). Therefore, $G \neq G_{\text{Kubo}}, G_{\text{Izuyama}}$ in such a case. Hence, to obtain the correct value of G by the Kubo formula, one must find the relation between $\Delta\phi_{\text{ext}}$ and $\Delta\mu$. Unfortunately, no systematic way of doing this has been developed.

The same can be said for mesoscopic conductors, Fig. 2(b), for which $e\Delta\phi \neq \Delta\mu$ in general if one takes the differences between both ends of the QWR. Therefore, $G \neq G_{\text{Kubo}}, G_{\text{Izuyama}}$. It is only for fortunate cases that G_{Kubo} or G_{Izuyama}

coincides with G. For example, Kawabata [49] calculated G_{Izuyama} for the case where the backward scattering with amplitude $V(2k_{\text{F}})$ is present, which had been neglected in the previous calculations. He found that

$$G_{\text{Izuyama}} = \frac{e^2}{\pi\hbar}\left[1 + \frac{V(2k_{\text{F}})}{2\pi\hbar v_{\text{F}}}\right]. \tag{46}$$

However, this result disagrees with G obtained in the previous sections. The origin of this discrepancy may be understood as follows. By taking the Fourier transforms of both sides of Eq. (18), we can see that only the $q \simeq 0$ component of the two-body potential v contributes to the screening of the electrostatic potential. On the other hand, both $q \simeq 0$ (forward) and $q \simeq 2k_{\text{F}}$ (backward) components of v contribute the Landau parameter f_{+-}, i.e., $f_{+-} = f_{+-}^{\text{forward}} + f_{+-}^{\text{backward}}$. Therefore, Eq. (9) shows that $\Delta\mu$ has a term (proportional to f_{+-}^{backward}) which cannot be interpreted as coming from the screening of ϕ_{ext}. If we interpret this term in terms of nonequilibrium thermodynamics (although it is not fully applicable because the local equilibrium is not established), the term may be interpreted as a *non-mechanical force* in Eq. (5):

$$\Delta\mu_{\text{c}} \to -(\hbar q/\pi)f_{+-}^{\text{backward}} + \cdots. \tag{47}$$

Here, $\cdots$ accounts for possible contributions from f_{++} and/or f_{++}^{forward}. Since $q \propto \langle I \rangle$, so is $\Delta\mu_{\text{c}}$. This means that a finite current $\langle I \rangle$ induces a finite non-mechanical force $\Delta\mu_{\text{c}}$, and $\langle I \rangle$ is driven by *both* $e\Delta\phi$ and $\Delta\mu_{\text{c}}$ in the steady state. Hence, $G \equiv \langle I \rangle/(\Delta\mu/e)$ is *not* equal to either $G_{\text{Kubo}} \equiv \langle I \rangle/\Delta\phi_{\text{ext}}$ or $G_{\text{Izuyama}} \equiv \langle I \rangle/\Delta\phi$. Therefore, the Kubo formula cannot give the correct value of G if the $q \simeq 2k_{\text{F}}$ component of the two-body potential is non-negligible, even if the screening of $\Delta\phi_{\text{ext}}$ is correctly taken into account, because a non-mechanical force is inevitably induced. Note that this is not the unique problem of the Kubo formula, but *a common problem of many microscopic theories* which calculate a nonequilibrium state by applying a mechanical force.

A possible way of getting the correct result by the Kubo formula would be to apply the formula to a larger system that includes the homogeneous reservoirs or leads [50]: in that case, $e\Delta\phi = \Delta\mu$, as sketched in Fig. 2, and thus $G = G_{\text{Izuyama}}$. However, this seems very difficult because it is almost equivalent to trying to solve the Schrödinger equation of the total system, including complicated processes that lead to the mixing property and to the equality $e\Delta\phi = \Delta\mu$. Another possible solution may be to apply Zubarev's method [2], which, to the authors' knowledge, has not been applied to interacting electrons in mesoscopic conductors. However, one must also include (a part of) reservoirs into the Hamiltonian because Zubarev's method assumes that macroscopic variables (such as μ) are well-defined in the nonequilibrium steady state. As compared with these approaches, the formulations presented in sections 3 and 4 would be simpler ways of getting correct results which include effects of non-mechanical forces.

6   Deviation from the Quantized Conductance

For a clean QWR, we have obtained the quantized value $G = e^2/\pi\hbar$ in both cases of the FL and the TLL in sections 3.1, 3.2 and 4.6, using different formulations. The essential assumptions leading to this result are the following. (i) The QWR is clean enough and the temperature is low enough (zero temperature has been assumed for simplicity), so that scatterings by impurities, defects or phonons are negligible and e-e interactions are the only scattering mechanism. (ii) The boundaries between the QWR and reservoirs are smooth and slowly-varying so that reflections at the boundaries are absent. (iii) The reservoirs are large enough, so that they remain at equilibrium even in the presence of a finite current between the reservoirs through the QWR.

When some of these assumptions are not satisfied the observed conductance may deviate from the quantized value. For example, if boundary reflections are non-negligible, the transmittance T (calculated from the single-body Schrödinger equation) between the reservoirs through the QWR is reduced. This results in the reduction of G by the factor T for non-interacting electrons. For interacting electrons, G will be further reduced for the TLL because the TLL will be "pinned" by the reflection potential at the boundaries. This can be understood simply as follows: Although the TLL of infinite length is a liquid, for which a long-range order is absent, it behaves like a solid at a short distance. Hence, the TLL is pinned by a local potential, like a charge-density wave is. This is the physical origin of the vanishing G (at zero temperature) for the case where a potential barrier is located *in* the TLL [4,5]. Since the pinning occurs irrespective of the position of the local potential, the TTL would be pinned also by the boundary reflections. Note that if one neglects the weakening of v_{11} in reservoirs (due to the broadening of $W(x)$, as shown in section 4.2), the TTL would then be pinned also by impurities in reservoirs [50].

Another example is dissipation by, say, phonon emission. By the dissipation, the 1d system will lose any correlations over a distance L_{rlx}, where L_{rlx} is the "maximal energy relaxation length" [7,34], which is generally longer than the simple dephasing length (over which an energy correlation may be able to survive). In such a case the 1d system of length L ($> L_{rlx}$) will behave as a series of independent conductors of length L_{rlx}. One will then observe Ohm's law [21]:

$$G_{obs} \simeq (L_{rlx}/L) \times (e^2/\pi\hbar). \tag{48}$$

Acknowledgment The authors are grateful to helpful discussions with A. Kawabata, T. Arimitsu, and K. Kitahara. This work has been supported by the Core Research for Evolutional Science and Technology (CREST) of the Japan Science and Technology Corporation (JST).

References

1. See, e.g., F. B. Callen, *Thermodynamics and Introduction to Thermostatistics, 2nd ed.* (Wiley, New York 1985) Chapter 14

2. D. Zubarev, *Nonequilibrium Statistical Thermodynamics* (Plenum, New York 1974)
3. W. Apel and T.M. Rice, Phys. Rev. B **26**, 7063 (1982)
4. C. L. Kane and M. P. A. Fisher, Phys. Rev. Lett. **68**, 1220 (1992)
5. A. Furusaki and N. Nagaosa, Phys. Rev. B **47**, 4631 (1993)
6. M. Ogata and H. Fukuyama, Phys. Rev. Lett. **73**, 468 (1994)
7. A. Shimizu and M. Ueda, Phys. Rev. Lett. **69**, 1403 (1992)
8. D.L. Maslov and M. Stone, Phys. Rev. B **52**, R5539 (1995) D.L. Maslov, another chapter of this book.
9. V.V. Ponomarenko, Phys. Rev. B **52**, R8666 (1995)
10. A. Kawabata, J. Phys. Soc. Jpn. **65**, 30 (1996)
11. I. Safi and H. J. Schulz, Phys. Rev. **B52**, R17040 (1995)
12. Y. Oreg and A. Finkel'stein, Phys. Rev. **B54**, R14265 (1996)
13. The role of the factor $1/e$ is just to convert the unit of $\Delta\mu$ into volt.
14. R. Landauer, IBM J. Res. Dev. **32**, 306 (1988) M. Büttiker, *ibid*, 317
15. It is sometimes argued that for a system with the translational symmetry the e-e interaction would not affect the value of G. For a mesoscopic conductor, however, $\Delta\mu$ (hence G) can be defined only when 2d or 3d reservoirs are connected, which violates the translational symmetry.
16. S. Tomonaga, Prog. Theor. Phys. **5**, 544 (1950)
17. J.M. Luttinger, J. Math. Phys. **4**, 1154 (1963)
18. F.D.M. Haldane, J. Phys. C **14**, 2585 (1981)
19. N. Kawakami and S.-K. Yang, J. Phys. **C3**, 5983 (1991)
20. S. Tarucha, T. Honda and T. Saku, Solid State Commun. **94**, 413 (1995)
21. A. Shimizu, J. Phys. Soc. Jpn. **65**, 1162 (1996)
22. R. Kubo, J. Phys. Soc. Jpn. **12**, 570 (1957)
23. R. Kubo, M. Toda, N. Hashitsume, *Statistical Physics II*, 2nd ed. (Springer, New York 1995) Chapter 4
24. M. Suzuki, Physica **51**, 277 (1971)
25. G. D. Mahan, *Many-Particle Physics*, 2nd ed. (Plenum, New York 1990) section 3.8A
26. E. Ott, *Chaos in Dynamical Systems* (Cambridge Univ., Cambridge 1993)
27. H. Nakano and M. Hattori, *What is the ergodicity?* (Maruzen, Tokyo, 1994) [in Japanese]
28. M. Pollicott and M. Yuri, *Dynamical Systems and Ergodic Theory* (Cambridge Univ., Cambridge 1998)
29. A. Shimizu, unpublished.
30. In thermodynamics, μ is called an "electrochemical potential", whereas μ_c is called a chemical potential [1]. We here follow the terminology used in the solid state physics, where μ is called a chemical potential [3–12,14,21]. Note that it is not $\Delta\phi$ but $\Delta\mu$ that is applied by a battery.
31. T. Izuyama, Prog. Theor. Phys. **25**, 964 (1961)
32. For example, the $\nabla\mu_c$ term is *dominant* in p-n junctions.
33. Gradients or differences of density, temperature, and so on, are driving forces that induce a finite current. These driving forces are called "non-mechanical forces" because *they cannot be represented as a mechanical term in the Hamiltonian* [2]. This point is sometimes disregarded in the literature.
34. A. Shimizu, M. Ueda and H. Sakaki, Jpn. J. App. Phys. Series 9, 189 (1993)
35. G.B. Lesovik, JETP Lett. **49**, 514 (1989)
36. M. Büttiker, Phys. Rev. Lett. **65**, 2901 (1990)
37. Y.P. Li, D.C. Tsui, J.J. Hermans, J.A. Simmons and G. Weimann, Appl. Phys. Lett. **57**, 774 (1990)

38. F. Liefrink, J.I. Dijkhuis, M.J.M. de Jong, L.W. Molenkamp and H. van Houten, Phys. Rev. B **49**, 14066 (1994)

39. The only exception is the NEN of (tunnel) junctions with a high barrier. In this case one can take the approximate equilibrium state with $\Delta\mu > 0$ as the equilibrium state of the Kubo formula.

40. R. Haag, *Local Quantum Physics* 2nd ed. (Springer, New York 1996).

41. P. Nozières, *Theory of Interacting Fermi Systems* (W. A. Benjamin, Amsterdam 1964) Chapter 1

42. This relation is derived from the identity, $\langle I \rangle = e\langle P \rangle / m\mathcal{L}$, where $\langle I \rangle$ and $\langle P \rangle$ denote the DC components of the current and the total momentum, respectively. *Remark:* This identity does not hold for the TLL if one describes it by the Luttinger model [17] because of its idealized linear dispersion. As a result, the mechanisms leading to the universal value of G are different between the FL and TLL, as described in A. Shimizu, J. Phys. Soc. Jpn. **65**, 3096 (1996).

43. Note that this equality should be satisfied only under the condition (which we assume) that the boundary reflections are negligible.

44. A. Shimizu and T. Miyadera, Physica B **249-251**, 518 (1998)

45. H. Kato and A. Shimizu, unpublished

46. This is an adiabatic approximation, which has been widely used in studies of optical waveguides. The equation for the optical field in a waveguide has the same form as the single-body Schrödinger equation of an electron in a quantum wire.

47. Note that the second-order treatment of $\hat{V}_{1\mathrm{R}}$ does not limit the accuracy of the theory because, e.g., the result for the steady current does not include $\hat{V}_{1\mathrm{R}}$. Only the accuracy of the relaxation time γ is limited to the second order.

48. Precisely speaking, the equality $\hat{\zeta}_1(t') \simeq \hat{\zeta}_1(t)$ holds not for $\hat{\zeta}_1$ but for expectation values evaluated from $\hat{\zeta}_1$. We therefore apply Eq. (34) to expectation values, as demonstrated in sections 4.5 and 4.6.

49. A. Kawabata, unpublished.

50. For an approach similar (but different) to this, see A. Kawabata, J. Phys. Soc. Jpn. **67**, 2430 (1998) and A. Kawabata, in the next chapter of this book.

A Linear Response Theory of 1D-Electron Transport Based on Landauer Type Model

Arisato Kawabata

Department of Physics, Gakushuin University
Mejiro, Toshima-ku Tokyo 171-8588, Japan

Abstract. Effects of the interaction on the electron transport in one-dimensional systems are discussed on a model like that of Landauer's theory. The model consists of one-dimensional channel with electron reservoirs at its ends. As the driving force of the current, we assume a chemical potential difference between the reservoirs, and we apply Kubo's linear response theory to calculate the current. As for the effects of interaction, it is shown that the conductance at 0K is not affected by the interaction if the electrons in 1D-channel behave like Fermi liquid. The case of Tomonaga-Luttinger liquid is not simple. Within the present model, the conductance depends on the length l of 1D-channel like $l^{-\gamma}$, where γ is a positive constant dependent on the strength of the interaction.

1 Introduction

Electron transport in one-dimensional interacting electron systems is not yet fully understood in spite of its simplicity. Even in the absence of the potential scattering, until recently it has been believed that the conductance is renormalized by the electron–electron interaction like $2Ke^2/h$, where $0 < K < 1$ for a repulsive interaction [1]. The experiments by Tarucha et al. [2], however, indicate that the renormalization is absent.

An explanation for the absence of the renormalization was given by the author: As long as the conductance is concerned, L–T (Tomonaga–Luttinger) liquid theory is equivalent to the self-consistent theory. The apparent renormalization is due to the wrong definition of the conductance, and the renormalization can be got rid of by defining the conductance correctly with respect to the renormalized electric field [3], according to the remark by Izuyama [4]. Maslov and Stone, Ponomarenko, and Safi and Schulz also proposed a theory based on a different model.

All the theories mentioned above are based on models of an infinitely long one-dimensional system with an electric field as a current driving force. On the other hand, the systems used in the experiments are rather similar to Landauer's model in his theory of electron transport [8–10]. Moreover, within one-dimensional model, we need rather delicate limiting processes to obtain a meaningful result for the static conductance from a dynamical conductance. Therefore, it is important to investigate the interaction effects using Landauer type model.

The purpose of this paper is to develop a theory of electron transport in one-dimension based on Landauer type model. We consider a model of one-dimensional channel of finite length with reservoirs attached to its ends. As the current driving force we assume a chemical potential difference between the reservoirs. We will apply Kubo's linear response theory [11] to calculate the current. First we will investigate the case of non-interacting electrons, and then we will extend the theory to interacting electrons.

2   Landauer Type Model

A somewhat realistic model of one-dimensional system is shown in Fig. 1: Two-dimensional electrons in a $x - y$ plane are confined between the shaded regions by an electrostatic potential. The part of the system bounded by the parallel boundaries will be called a '1D channel', and the parts attached to that will be called 'reservoirs'. Here we take the x-axis along the 1D channel. In a 1D channel, the solutions of the Schrödinger equation for a given energy are of the form $e^{ip_n x}\phi_n(y)$, where n is a quantum number associate with the freedom along y-axis and p_n is real or complex. For an appropriate energy, p_n is real only for one value of n, e.g., $n = 0$. The eigenstate of the whole system which is connected to this state in 1D channel will be called '1D state'. The amplitude of other eigenstates decreases rapidly in 1D channel, and they will be called 'reservoir states'. For the details of the separation of those states, the reader is referred to ref. [12].

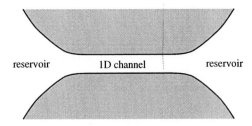

reservoir   1D channel   reservoir

Fig. 1. A realistic model of one-dimensional system connected to reservoirs

In this paper we consider a model in which the exchange of electrons between 1D state and the reservoir states are taken into account in the simplest way: We approximate 1D channel with pure one-dimensional system along x-axis which spatially overlaps with the reservoirs in the regions $l/2 < x$ and $x < -l/2$ (see Fig. 2). We introduce impurity scattering as the mechanism of the exchange of electrons between 1D states and the reservoir states.

For the moment we neglect the spin degeneracy. The hamiltonian of the system consists of $\mathcal{H}_1$ for the electrons in 1D states, $\mathcal{H}_L(\mathcal{H}_R)$ for the electrons in the reservoir states in L(R), and $\mathcal{H}_{iL}(\mathcal{H}_{iR})$ for impurity scattering in the reservoir L(R):

$$\mathcal{H} = \mathcal{H}_1 + \mathcal{H}_L + \mathcal{H}_R + \mathcal{H}_{iL} + \mathcal{H}_{iR} . \tag{1}$$

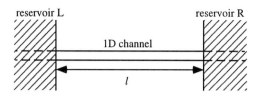

Fig. 2. The model system: The 1D channels connected to the reservoirs R and L

The each hamiltonian assumes the following forms:

$$\mathcal{H}_1 = \int \frac{dp}{2\pi} \varepsilon_p a_p^\dagger a_p \,, \tag{2}$$

where $\varepsilon_p = \hbar^2 p^2 / 2m - \mu$, μ being the chemical potential, and a_p is the annihilation operator of an electron of wave number p along the channel, and

$$\mathcal{H}_L = \sum_k E_k b_k^\dagger b_k \,, \tag{3}$$

where b_k is the annihilation operator of an electron in a reservoir state specified by the quantum number k , which is not necessarily the wave number. Here the chemical potential is included in the eigenenergy E_k. $\mathcal{H}_R$ is defined in the same way.

As for the impurity scattering, it's role is to inject electrons to 1D states from the reservoir states and vise versa. Hence we neglect the scatterings within 1D states and those within the reservoir states. Therefore we put

$$\mathcal{H}_{iL} = \int_{-\infty}^{-l/2} \{\psi^i(x)\Psi_L(x) + \psi(x)\Psi_L^\dagger(x)\} V_L(x) dx \,, \tag{4}$$

where $\psi(x)$ is the field operator of electrons in 1D states and $\Psi_L(\mathbf{r})$ is those in the reservoir state in L:

$$\psi(x) = \int_{-\infty}^{\infty} \frac{dp}{2\pi} e^{ipx} a_p \,, \tag{5}$$

$$\Psi_L(\mathbf{r}) = \sum_k B_k(\mathbf{r}) b_k \,, \tag{6}$$

with $x = (x, 0)$, $B_k(\mathbf{r})$ being the wave functions of a reservoir states in L. The impurity potential $V_L(x)$ is of the form

$$V_L(x) = V_0 \sum_i \delta(x - x_i) \,, \tag{7}$$

where the impurities are uniformly distributed over the region $x_i < -l/2$. The hamiltonian $\mathcal{H}_{iR}$ is defined in the same way.

3   Linear Response Theory

As the current driving force, we introduce the chemical potential difference between the reservoirs L and R. It is described by the hamiltonian

$$\mathcal{H}' = \frac{\Delta\mu}{2}(N_{\mathrm{L}} - N_{\mathrm{R}}),\tag{8}$$

where N_{L} is the total numbers of electrons in the reservoirs L:

$$N_{\mathrm{L}} = \int_{\mathrm{L}} \Psi_{\mathrm{L}}^{\dagger}(\mathbf{r})\Psi_{\mathrm{L}}(\mathbf{r})\,\mathrm{d}\mathbf{r},\tag{9}$$

and $\int_{\mathrm{L}} \mathrm{d}\mathbf{r}$ means that the integral is to be done over the region $x < -l/2$. N_{R}, the total numbers of electrons in the reservoirs R, is defined in the same way.

If $\Delta\mu > 0$, the electrons in the reservoir states in L are injected into 1D states and corrected at the reservoir R. Therefore, the current in 1D channel is given by

$$I = e\left\langle\frac{\mathrm{d}N_{\mathrm{R}}}{\mathrm{d}t}\right\rangle = -e\left\langle\frac{\mathrm{d}N_{\mathrm{L}}}{\mathrm{d}t}\right\rangle,\tag{10}$$

e being the absolute value of the electronic charge.

We calculate the current to the first order in $\Delta\mu$. Using the linear response theory [11], we find that

$$I = -\lim_{\delta\to+0}\frac{ie\Delta\mu}{2\hbar}\int_0^{\infty}\left\langle\left[\frac{\mathrm{d}N_{\mathrm{R}}(t)}{\mathrm{d}t}, (N_{\mathrm{R}} - N_{\mathrm{L}})\right]\right\rangle e^{-\delta t}\,\mathrm{d}t,\tag{11}$$

where $\langle\cdots\rangle$ means the thermal average $\mathrm{Tr}\{e^{-\beta\mathcal{H}}\cdots\}/\mathrm{Tr}\{e^{-\beta\mathcal{H}}\}$. Therefore, from eq. (10) it follows that the conductance is given by

$$\begin{aligned}G &= \frac{I}{\Delta\mu/e}\\ &= \lim_{\delta\to+0}\frac{ie^2}{\hbar}\int_0^{\infty}\left\langle\left[\frac{\mathrm{d}N_{\mathrm{R}}(t)}{\mathrm{d}t}, N_{\mathrm{L}}\right]\right\rangle e^{-\delta t}\,\mathrm{d}t.\end{aligned}\tag{12}$$

Here we define a retarded Green's function

$$D(\omega) = -\lim_{\delta\to+0}\mathrm{i}\int_0^{\infty}\langle[N_{\mathrm{R}}(t), N_{\mathrm{L}}]\rangle e^{(\mathrm{i}\omega-\delta)t}\,\mathrm{d}t.\tag{13}$$

Then, integrating the right hand side of eq. (12) by part, we easily obtain

$$G = \lim_{\omega\to0}\mathrm{i}\omega\frac{e^2}{\hbar}D(\omega).\tag{14}$$

Thus the problem reduces to the calculation of the Green's function $D(\omega)$.

4   Thermal Green's Function Technique

4.1   Thermal Green's Function and Retarded Green's Function

In this paper we treat only the case of zero temperature, but in the following we will use the thermal Green's function formalism because it is an excellent method to treat the many-body interaction. [13,14]

The thermal Green's function corresponding to $D(\omega)$, is defined as

$$\mathcal{D}(\omega_n) \equiv -\int_0^{\beta\hbar} \langle \bar{N}_{\mathrm{R}}(\tau) N_{\mathrm{L}} \rangle e^{\mathrm{i}\omega_n \tau} \, \mathrm{d}\tau \,, \qquad (15)$$

where

$$\bar{N}_{\mathrm{R}}(\tau) = e^{\mathcal{H}\tau} N_{\mathrm{R}} e^{-\mathcal{H}\tau} \,, \qquad (16)$$

and ω_n is the Matsubara frequency, i.e., $\omega_n = 2\pi k_{\mathrm{B}} T n/\hbar$, n being an integer. $D(\omega)$ is obtained by an analytic continuation

$$D(\omega) = \lim_{\delta \to +0} \mathcal{D}(\omega_n \to -\mathrm{i}\omega + \delta) \,. \qquad (17)$$

We calculate the Green's function $\mathcal{D}(\omega_n)$ by Feynman graph method [14]. We will treat the impurity-scattering Hamiltonian

$$\mathcal{H}_{\mathrm{i}} \equiv \mathcal{H}_{\mathrm{iL}} + \mathcal{H}_{\mathrm{iR}} \,, \qquad (18)$$

as a perturbation.

In addition to the thermal average, the Green's functions have to be averaged over the positions of the impurities. We take the limit of weak impurity potential V_0 and high impurity density c so that cV_0^2 is finite. Then the correlation function of the potential is given by

$$\langle V(x)V(x') \rangle_{\mathrm{imp}} = cV_0^2 \delta(x - x') \,. \qquad (19)$$

As for $\langle V(x) \rangle_{\mathrm{imp}}$, which is divergent in the above mentioned limit, it can be absorbed in the chemical potential. Then the correlation functions of odd order vanish, and those of even order can be decomposed into the products of 2nd order correlation functions with all the possible combinations of the arguments.

In Fig. 3 we show the simplest Feynman graph which describes the process of electron transfer between the reservoirs L and R. Here the thin solid lines represent the the Green's functions of electrons in 1D channel $\mathcal{G}_0(x, x'; \varepsilon_n)$, and the thick solid lines represent those of the reservoir states $\mathcal{G}_{\mathrm{L}}(\mathbf{r}, \mathbf{r}'; \varepsilon_n)$ and $\mathcal{G}_{\mathrm{R}}(\mathbf{r}, \mathbf{r}'; \varepsilon_n)$. The dotted lines represents the impurity potential given by eq. (19).

The Green's functions are defined as

$$\mathcal{G}_0(x, x'; \varepsilon_n) = -\int_0^{\beta\hbar} \langle \psi(x, \tau)\psi^\dagger(x') \rangle_0 e^{\mathrm{i}\varepsilon_n \tau} \mathrm{d}\tau \,, \qquad (20a)$$

$$\mathcal{G}_{\mathrm{L}}(\mathbf{r}, \mathbf{r}'; \varepsilon_n) = -\int_0^{\beta\hbar} \langle \Psi_{\mathrm{L}}(\mathbf{r}, \tau)\Psi_L^\dagger(\mathbf{r}') \rangle_0 e^{\mathrm{i}\varepsilon_n \tau} \mathrm{d}\tau \,, \qquad (20b)$$

$$\mathcal{G}_{\mathrm{R}}(\mathbf{r}, \mathbf{r}'; \varepsilon_n) = -\int_0^{\beta\hbar} \langle \Psi_{\mathrm{R}}(\mathbf{r}, \tau)\Psi_R^\dagger(\mathbf{r}') \rangle_0 e^{\mathrm{i}\varepsilon_n \tau} \mathrm{d}\tau \,, \qquad (20c)$$

where $\varepsilon_n = \pi k_B T (2n+1)/\hbar$, n being an integer.

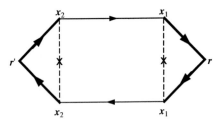

Fig. 3. The simplest Feynman graph for the electron transfer between the reservoirs L and R

The contribution of this graph to $\mathcal{D}(\omega_n)$ is

$$\mathcal{D}(\omega_n) = -\frac{(cV_0^2)^2}{\hbar^4}\frac{k_B T}{\hbar}\sum_{\varepsilon_n}\int_R d\mathbf{r}d\mathbf{x}_1\int_L d\mathbf{r}'d\mathbf{x}_2\ \mathcal{G}_R(\mathbf{x}_1, \mathbf{r}; \varepsilon_n)$$
$$\times\ \mathcal{G}_R(\mathbf{r}, \mathbf{x}_1; \varepsilon_n + \omega_n)\mathcal{G}_0(\mathbf{x}_2, \mathbf{x}_1; \varepsilon_n)\mathcal{G}_0(\mathbf{x}_1, \mathbf{x}_2; \varepsilon_n + \omega_n)$$
$$\times\ \mathcal{G}_L(\mathbf{r}', \mathbf{x}_2; \varepsilon_n)\mathcal{G}_L(\mathbf{x}_2, \mathbf{r}'; \varepsilon_n + \omega_n)\,, \tag{21}$$

where $\mathbf{x}_1 = (x_1, 0)$ and $\mathbf{x}_2 = (x_2, 0)$.

From eq. (6) and eq. (20b) we find that

$$\mathcal{G}_L(\mathbf{r}, \mathbf{r}'; \varepsilon_n) = \sum_k \frac{\hbar B_k(\mathbf{r}) B_k^*(\mathbf{r}')}{i\hbar\varepsilon_n - E_k}\,. \tag{22}$$

Then from the orthogonality of $B_k(\mathbf{r})$'s it follows that

$$\int_L \mathcal{G}_L(\mathbf{r}', \mathbf{x}_2; \varepsilon_n)\mathcal{G}_L(\mathbf{x}_2, \mathbf{r}'; \varepsilon_n + \omega_n)\,d\mathbf{r}'$$
$$= \sum_k \frac{\hbar^2 B_k^*(\mathbf{x}_2) B_k(\mathbf{x}_2)}{\{i\hbar\varepsilon_n - E_k\}\{i\hbar(\varepsilon_n + \omega_n) - E_k\}}\,. \tag{23}$$

As is seen from eq. (14), in order to obtain a finite value for the conductance, $\mathcal{D}(\omega_n)$ must be proportional to $1/\omega$, after the analytic continuation $\omega_n \to -i\omega + \delta$ is made. In fact, such singular behavior of $\mathcal{D}(\omega_n)$ can arise from the right hand side of eq. (23), because it is divergent in the limit $\varepsilon_n, \omega_n \to 0$. The important contribution to this divergence comes from $E_k \sim 0$. Here we assume that the contribution of the states with $E_k \sim 0$ to the electron density is spatially uniform. Namely, we define

$$F(E) \equiv \sum_k B_k^*(\mathbf{r}) B_k(\mathbf{r})\delta(E - E_k)\,, \tag{24}$$

and we neglect the dependence of $F(E)$ on $\mathbf{r}$. Then the right hand side of eq. (23) can be written in the form

$$\int \frac{\hbar F(E)}{i\omega_n} \left[\frac{1}{i\hbar\varepsilon_n - E} - \frac{1}{i\hbar(\varepsilon_n + \omega_n) - E} \right] dE , \qquad (25)$$

and for very small but finite ε_n and ω_n, we find that

$$\int_L \mathcal{G}_L(\mathbf{r}', \boldsymbol{x}_2; \varepsilon_n) \mathcal{G}_L(\boldsymbol{x}_2, \mathbf{r}'; \varepsilon_n + \omega_n) \, d\mathbf{r}'$$
$$= \begin{cases} 2\pi\hbar F(0)/\omega_n , & (\varepsilon_n(\varepsilon_n + \omega_n) < 0) \\ 0 , & (\varepsilon_n(\varepsilon_n + \omega_n) > 0) \end{cases} \qquad (26)$$

where we neglect the term which is not singular for $\omega_n \to 0$.

Thus we can do the integrals over $\mathbf{r}$ and $\mathbf{r}'$ in eq. (21), and for $\omega_n > 0$ we obtain

$$\mathcal{D}(\omega_n) = - \left(\frac{2\pi c V_0^2 F(0)}{\hbar \omega_n} \right)^2 \frac{k_B T}{\hbar} \sum_{-\omega_n < \varepsilon_n < 0}$$
$$\times \int_R d\boldsymbol{x}_1 \int_L d\boldsymbol{x}_2 \, \mathcal{G}_0(\boldsymbol{x}_2, \boldsymbol{x}_1; \varepsilon_n) \mathcal{G}_0(\boldsymbol{x}_1, \boldsymbol{x}_2; \varepsilon_n + \omega_n). \qquad (27)$$

The Green's function $\mathcal{G}_0(x, x'; \varepsilon_n)$ for a small positive value of ε_n is given by

$$\mathcal{G}_0(x, x'; \pm\varepsilon_n) = \mp \frac{i}{v_F} e^{(\pm i p_F - \varepsilon_n/v_F)|x-x'|} , \qquad (28)$$

where p_F and v_F are the Fermi wave number and the Fermi velocity, respectively.

We easily find that the integrals in eq. (27) diverge if we put eq. (28) into it, and to be consistent we have to calculate $\mathcal{G}_0(x, x'; \varepsilon_n)$ also in the lowest order in the impurity scattering.

4.2 Calculation of the Green's Function in 1D Channel

The Green's function $\mathcal{G}(x, x'; \varepsilon_n)$ of 1D states with impurity-scattering correction can be obtained by solving the equation

$$\mathcal{G}(x, x'; \varepsilon_n) = \mathcal{G}_0(x, x'; \varepsilon_n) + \int \mathcal{G}_0(x, x_1; \varepsilon_n)$$
$$\times \Sigma(x_1, x_2; \varepsilon_n) \mathcal{G}(x_2, x'; \varepsilon_n) \, dx_1 \, dx_2 , \qquad (29)$$

where $\Sigma(x_1, x_2; \varepsilon_n)$ is the self-energy part (divided by $\hbar$). As was mentioned in the above, we calculate $\Sigma(x_1, x_2; \varepsilon_n)$ to the lowest order in the impurity scattering. It vanishes unless x_1 and x_2 are in the same reservoirs after averaged

Fig. 4. The Feynman for the self-energy of the Green's function of 1D states

over the positions of the impurities. When x_1 and x_2 are in the reservoir L, it is of the form

$$\Sigma(x_1, x_2; \varepsilon_n) = \frac{cV_0^2}{\hbar^2}\delta(x_1 - x_2)\mathcal{G}_L(x_1, x_2; \varepsilon_n).\tag{30}$$

The Feynman graph for the self-energy is shown in Fig. 4.

As was mentioned in the above, the important contribution to $\mathcal{D}(\omega_n)$ comes from very small ε_n. Then, from eqs. (20b) and (24) we obtain

$$\Sigma(x_1, x_2; \varepsilon_n) = -i\frac{\pi cV_0^2 F(0)}{\hbar}\mathrm{sgn}(\varepsilon_n)\delta(x_1 - x_2).\tag{31}$$

We neglect the real part of the self-energy part, for its dependence on ε_n is weak and it can be included in the chemical potential.

Then eq. (29) becomes

$$\mathcal{G}(x, x'; \varepsilon_n) = \mathcal{G}_0(x, x'; \varepsilon_n)$$
$$-i\,\mathrm{sgn}(\varepsilon_n)\Gamma\int_{l/2\leq|x''|}\mathcal{G}_0(x, x'', \varepsilon_n)\mathcal{G}(x'', x', \varepsilon_n)\,\mathrm{d}x'',\tag{32}$$

with

$$\Gamma = \frac{\pi cV_0^2 F(0)}{\hbar}.\tag{33}$$

From eqs. (20a) and (32), we easily find that

$$\mathcal{G}(x, x', \varepsilon_n) = \mathcal{G}(x', x, \varepsilon_n),\tag{34}$$

and hereafter we will consider the case $x' < x$.

We assume that Γ is small enough so that the behavior of $\mathcal{G}(x, x', \varepsilon_n)$ is not very much different from that of $\mathcal{G}_0(x, x', \varepsilon_n)$. Then, in the regions $x < x''$ or $x'' < x'$, the integrand of the integral in eq. (32) oscillates rapidly. Therefore we will neglect the contributions from those regions.

Within this approximation we can solve eq. (32) assuming a form

$$\mathcal{G}(x, x'; \varepsilon_n) = \mathcal{G}_0(x, x', \varepsilon_n)\mathcal{F}(x, x').\tag{35}$$

As is seen from eq. (27) we need $\mathcal{G}(x, x'; \varepsilon_n)$ only for $x' < -l/2, l/2 < x$, and here we show the result in those regions:

$$\mathcal{F}(x, x') = e^{(l/2-x)/2\lambda} e^{(l/2+x')/2\lambda}, \tag{36}$$

where

$$\lambda = \frac{v_{\mathrm{F}}}{2\Gamma}. \tag{37}$$

Now we are ready to calculate the conductance.

5   Calculation of Conductance–Landauer Formula

We calculate $\mathcal{D}(\omega_n)$ by replacing $\mathcal{G}_0(x, x'; \varepsilon_n)$ with $\mathcal{G}(x, x'; \varepsilon_n)$ obtained above in eq. (27). The summation over ε_n in eq. (27) is limited in the region $-\omega_n < \varepsilon_n < 0$, and to calculate the conductance we take the limit of small frequency after the analytic continuation is made (see eqs. (14) and (17)). Therefore, the small values of ω_n are relevant, and we can replace ε_n and $\varepsilon_n + \omega_n$ in the Green's functions with infinitely small negative and positive frequencies, respectively.

Then from eq. (27) we easily find that

$$\mathcal{D}(\omega_n) = -\frac{2\pi}{\omega_n} \left(\frac{cV_0^2 F(0)}{\hbar} \right)^2$$
$$\times \int_{\mathrm{R}} \mathrm{d}x_1 \int_{\mathrm{L}} \mathrm{d}x_2 |\mathcal{G}(x_1, x_2; +0)|^2. \tag{38}$$

and from eqs. (33) and (35)-(37) that

$$\mathcal{D}(\omega_n) = -\frac{2\pi}{\omega_n} \left(\frac{cV_0^2 F(0)}{\hbar} \right)^2 \left(\frac{\lambda}{v_{\mathrm{F}}} \right)^2 = -\frac{1}{2\pi\omega_n}. \tag{39}$$

Doing the analytic continuation eq. (17) and putting it into eq. (14) we obtain the expression for the conductance. In order to take into account the spin degeneracy, which we have neglected so far, we only have to double $D(\omega)$. Then we have

$$G = \frac{e^2}{\pi\hbar}. \tag{40}$$

Thus we obtained Landauer formula for the case without potential scattering .

6   Effects of Electron–Electron Interaction

6.1   Vertex Corrections

Here we consider only the interaction between the electrons in 1D states. In this case $\mathcal{D}(\omega_n)$ consists in two parts. The first part, to be called $\mathcal{D}_1(\omega_n)$, is

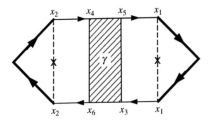

Fig. 5. The Feynman graph for $\mathcal{D}(\omega_n)$ with vertex correction

the same as $\mathcal{D}(\omega_n)$ in §4 except that the interaction is fully taken into account in $\mathcal{G}(x, x'; \varepsilon_n)$. The other part, $\mathcal{D}_2(\omega_n)$, is the one with a vertex correction, for which the Feynman graph is shown in Fig. 5. We can show that $\mathcal{D}_2(\omega_n)$ does not contribute to the conductance. For the details the reader is referred to Ref. [15].

Therefore only $\mathcal{D}_1(\omega_n)$ contributes to the conductance, and, even in the presence of the interaction, the process of deriving eq. (27) from eq. (21) does not affected by it, and (38) is valid if the interaction and the impurity scattering are taken into account in $\mathcal{G}(x, x'; \varepsilon_n)$.

Note that the conductance is determined only by the one-electron Green's function. It is one of the important outcomes of the present theory.

Below we will see how the conductance is affected by the interaction.

6.2   Conductance of 1D Fermi Liquid

Although it is widely accepted that interacting 1D electron system is not a Fermi liquid, it is instructive to apply the formula obtained above to Fermi liquid. At zero temperature, one-electron Green's function of Fermi liquid in momentum representation is of the form [14]

$$\mathcal{G}_0(p, \varepsilon_n) = \frac{a}{i\varepsilon_n - v^*(|p| - p_F)}, \tag{41}$$

for $p \sim p_F$ and $\varepsilon_n \sim 0$, where a is a positive constant and v^* is the renormalized Fermi velocity.

As we have seen in §5, we need the Green's function only in the limit $\varepsilon_n \to 0$. Therefore, when $|x-x'| \gg 1/p_F$, the Green's function in real space representation is given by

$$\mathcal{G}_0(x, x'; \pm 0) = \int_{-\infty}^{\infty} \frac{dp}{2\pi} \frac{ae^{ip(x-x')}}{\pm i0 - v^*(|p| - p_F)}$$

$$= \mp i\frac{a}{v^*} e^{\pm ip_F|x-x'|}, \tag{42}$$

because the main contribution to the integral comes from the regions $|p| \sim p_F$. Thus the calculation of the conductance reduces to that of non-interacting electron by replacing v_F with v^*/a. In fact, if we assume eq. (35) we easily find that it satisfies eq. (32) with $\lambda = v^*/(2a\Gamma)$. Here if the impurity potential is weak enough and $\lambda \gg p_F$, the regions $|x - x'| \sim \lambda \gg p_F$ mainly contribute to the integral in eq. (32). In addition it is the case also in the integral in eq. (38). Therefore the use of the approximate form of $G_0(x, x'; \varepsilon_n)$ for $|x - x'| \gg 1/p_F$ is justified.

As we have seen in eq. (40), the expression for the conductance does not contain v_F. Thus we find, as long as we assume the Fermi liquid, that the interaction gives no effects on the conductance.

6.3   Conductance of Tomonaga–Luttinger Liquid

The one-electron Green's function of T-L liquid is very much different from that of Fermi liquid: It does not have a simple pole as a function of ε_n even for low lying excitations. The details of the behavior of the Green's function depends on the form of the interaction potential. In the following, we will investigate the case treated by Luther and Peschel. [16]

In this case, U_p, the Fourier transform of the interaction potential, is defined in terms of two real positive parameters γ and r through a function $\varphi(p)$ as

$$U_p = -\pi v_F \tanh 2\varphi(p) , \tag{43a}$$

$$\sinh^2 \varphi(p) = \gamma e^{-|p|r} . \tag{43b}$$

In weak coupling cases, i.e., $\gamma \ll 1$, we easily find that $U_p = 2\pi v_F \sqrt{\gamma} e^{-|p|r/2}$, and that the interaction potential is a Lorentzian in the real space.

The retarded Green's function in space-time representation is given by

$$G_0(x, x'; t) = i\theta(t)\pi^{-1} e^{ip_F|x-x'|}$$

$$\times \mathrm{Im}\left[(|x - x'| - v^*t + i0)^{-1} \left\{ \left(\frac{|x-x'|}{r}\right)^2 + \left(1 + \frac{iv^*t}{r}\right)^2 \right\}^{-\gamma} \right] ,$$

$$\tag{44}$$

where $v^* = v_F \mathrm{sech}\, \varphi(0)$. The retarded Green's function $G_0(x, x'; \varepsilon)$ in frequency representation and the thermal Green's function $\mathcal{G}_0(x, x'; \varepsilon_n)$ agree with each other in the limit $\varepsilon \to 0, \varepsilon_n \to +0$, and it follows that

$$\lim_{\varepsilon_n \to +0} \mathcal{G}_0(x, x'; \varepsilon_n) = \int_0^\infty G_0(x, x'; t)\mathrm{d}t . \tag{45}$$

For $|x - x'| \gg r$, the main contribution to the above integral comes from the regions $v^*t \gtrsim |x - x'|$, and from the dimensional analysis we easily find that

$$\lim_{\varepsilon_n \to 0} \mathcal{G}_0(x, x'; \varepsilon_n) = i\frac{C}{v^*} e^{ip_F|x-x'|} \left(\frac{r}{|x - x'|}\right)^{2\gamma} , \tag{46}$$

C being a numerical constant. The power law decay of the amplitude of the Green's function is one of the general properties of T-L liquid. [1]

The equations eq. (32), and eq. (38) are valid also for T-L liquid. It is not easy to solve eq. (32), but, if Γ is small enough, the behavior of $\mathcal{G}(x, x'; \varepsilon_n)$ will not be very much different from that of $\mathcal{G}_0(x, x'; \varepsilon_n)$. Then, neglecting the contributions of rapidly oscillating term in the integral, from eq. (32) we find that $\mathcal{G}(x, x'; \varepsilon_n) = \mathcal{G}_0(x, x'; \varepsilon_n)$ for $-l/2 < x, x' < l/2$, and that $\mathcal{G}(l/2, -l/2; +0)\mathcal{G}(-l/2, l/2; -0) \sim (\frac{r}{l})^{4\gamma}/v^{*2}$. Therefore, within a crudest approximation, we estimate the conductance:

$$G \sim \frac{e^2}{\pi\hbar}\left(\frac{r}{l}\right)^{4\gamma}. \tag{47}$$

Thus, the present theory predict that the conductance of T-L liquid is much smaller than $e^2/\pi\hbar$ for sufficiently long 1D channel. It has been known that the conductance of T-L liquid in infinitely long 1D channel vanishes in the presence of potential scattering , however weak may it be. Since we have introduced impurity in the reservoir, the results obtained above is not surprising.

7   Summary and Discussion

We have formulated a Landauer type approach to electron conduction in 1D channel, explicitly taking into account the reservoirs in microscopic level. For non-interacting electrons, the present theory gives the same results as those obtained using the models of infinitely long one-dimensional systems. The present theory is, however, free from the delicate limiting processes.

One of the important outcome of the theory is that at zero temperature the conductance is expressed in terms of one-electron Green's function with zero frequency in 1D channel, even in the presence of electron-electron interaction. It is reasonable because we calculate the current as the number of electrons par time transferred from one of the reservoirs to the other: At zero temperature and for infinitesimally small chemical potential difference, the electrons are injected from the reservoir just onto the Fermi level of 1D channel and vice versa. This process can be described by the one-electron Green's function. On the other hand, Kubo formula consists in two-electron Green's functions. In most cases it is much easier to calculate one-electron Green's functions than to calculate two-electron Green's function, and it is the advantage of the present formalism.

As for the effects of the interaction, we found that the conductance of Fermi liquid is the same as that of non-interacting electron, while that of T-L liquid decreases by power low as the length of 1D channel increases. In the theories based on Kubo formula, the electrons are driven uniformly by the electric potential, and the current carrying state is the ground state in a moving frame. Hence the conductance is the same as that of non-interacting electrons. In the present theory the current carrying state is very much different. Suppose an electron is injected onto the Fermi level of the ground state of T-L liquid of N electrons. Then the resultant state is not the ground state but a linear combination of eigenstates of $N+1$ electrons in a moving frame. In fact, the density of states of T-L liquid is zero for $p = \pm p_F$ and $\omega = 0$, which means the amplitude of the ground state is zero. Thus it is reasonable that we have obtained an essentially different result for the conductance. It should be note that the quantized conductance $G \sim 2e^2/h$ is observed experimentally at rather high temperatures, [2] for which the present theory is not applicable.

The model system in this theory is somewhat different from the typical one-dimensional system like the one shown in Fig.1. We can find, however, real systems similar to our model: One of the examples is the system used in the experiments by Yacoby et al. [17]. Their sample is composed of 1D-channel and two-dimensional electrons attached to it, which play the role of the reservoirs (see Fig. 6). Another example is the experiments on carbon nanotube by Frank et al. [18]. They dipped one of the ends of nanotube in liquid mercury. At least this end is concerned the system is similar to our model. Thus we can expect to observe L-T liquid effects in such systems.

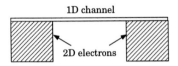

Fig. 6. A schematic view of the sample used in the experiments by Yacoby et al. [17]

Acknowledgments

The author is grateful to the organizers of the 219th WEH workshop for inviting him to it. This work is partly supported by 'High Technology Research Center Project' of Ministry of Education, Sciences, Sports and Culture.

References

1. J. Solyom: Adv. Phys. **28** 201, (1979)
2. S. Tarucha, T. Honda and T. Saku: Solid State Commun. **94** 413, (1995)
3. A. Kawabata: J. Phys. Soc. Jpn. **65** 30, (1996)
4. T. Izuyama: Prog. Theor. Phys. **25** 964, (1961)
5. D.L. Maslov and M. Stone: Phys. Rev. B **52** R5539, (1995)
6. V.V. Ponomarenko: Phys. Rev. B **52** R8666, (1995)
7. I. Safi and H. J. Schulz: Phys. Rev. B **52** R17040, (1995)
8. R. Landauer: IBM J. Res. & Dev. **1** 223, (1957)
9. R. Landauer: Philos. Mag. **21** 863, (1970)
10. M. Büttiker, Y. Imry, R. Landauer and S. Pinhas: Phys. Rev. B **31** 6207, (1985)
11. R. Kubo: J. Phys. Soc. Jpn. **12** 570, (1957)

12. A. Shimizu and T. Miyadera: to be published in Physica B (*Proc. Int. Conf. Electronic Properties of Two-Dimensional Systems* (1997, Tokyo))
13. T. Matsubara: Prog. Theor. Phys. **14** 351, (1955)
14. A.A. Abrikosov, L.P. Gor'kov and I.E. Dzyaloshinskii: *Methods of Quantum Field Theory in Statistical Physics* (Pergamon Press, London, 1963, and Dover Publication Inc., New York, 1975)
15. A. Kawabata: J. Phys. Soc. Jpn. **67** 2430, (1998)
16. A. Luther and I. Peschel: Phys. Rev. B **9** 2911, (1974)
17. A. Yacoby, H.L. Stormer, N.S. Wingreen, L.N. Pfeiffer, K.W. Baldwin and K.W. West: Phys. Rev. Lett. **77** 4612, (1996)
18. S. Frank, P. Poncharal, Z.L. Wang and W.A. De Heer: Science **280** 1744, (1998)

Gapped Phases of Quantum Wires

Oleg A. Starykh[1], Dmitrii L. Maslov[2], Wolfgang Häusler[3], and Leonid I. Glazman[4]

[1] Department of Applied Physics, Yale University, P. O. Box 208284
New Haven, CT 06520-8284
[2] Department of Physics, University of Florida, P. O. Box 118440
Gainesville, FL 32611-8440
[3] Institut für Theoretische Physik der Universität Hamburg
Jungiusstr. 9, D-20355 Hamburg, Germany
[4] School of Physics and Astronomy
University of Minnesota, Theoretical Physics Institute
116 Church St., SE Minneapolis, MN 55455

Abstract. We investigate possible nontrivial phases of a two-subband quantum wire. It is found that inter- and intra-subband interactions may drive the electron system of the wire into a gapped state. If the nominal electron densities in the two subbands are sufficiently close to each other, then the leading instability is the inter-subband charge-density wave (CDW). For larger difference in the densities, the interaction in the inter-subband Cooper channel may lead to a superconducting instability. The total charge density mode, responsible for the conductance of an ideal wire, always remains gapless, which enforces the two-terminal conductance to be at the universal value of $2e^2/h$ per occupied subband. On the contrary, the tunneling density of states (DOS) in the bulk of the wire acquires a hard gap, above which the DOS has a non-universal singularity. This singularity is weaker than the square-root divergency characteristic for non-interacting quasiparticles near a gap edge due to the "dressing" of massive modes by a gapless total charge density mode. The DOS for tunneling into the end of a wire in a CDW-gapped state, however, preserves the power-law behavior due to the frustration the edge introduces into the CDW order. This work is related to the vast literature on coupled 1D systems, and most of all, on two-leg Hubbard ladders. Whenever possible, we give derivations of the important results by other authors, adopted for the context of our study.

1 Introduction

From a theorist's point of view, electrons in quantum wires should provide a simplest realization of a Luttinger liquid [1,2]. Indeed, as the motion is confined in the direction transverse to the axis of a wire, the system is effectively one-dimensional; the electron-electron interaction is strong enough (typically, of the order of the Fermi energy) for the interaction effects not be washed by the temperature; and the state-of-the-art wires (at least the semiconductor version of them) are clean enough for disorder effects to be sufficiently weak. A clear experimental proof of the existence of the Luttinger-liquid state in quantum wires would be provided by tunneling into a wire, either in the middle or into the end. As is well-known, the tunneling density of states (DOS) of a Luttinger liquid reveals a pseudogap behavior, i.e, it is suppressed at energies close to the Fermi-energy, which should result in a sublinear (power-law) bias dependence of the tunneling current, and in a power-law temperature dependence of the Ohmic conductance. This finite-bias and finite-temperature measurements has already been performed on a very special realization of quantum wires – carbon nanotubes, and observed non-linear current-voltage dependences were interpreted in terms of the Luttinger-liquid theory [3,4]. Features of resonant tunneling, characteristic for a Luttinger liquid, have recently been observed on GaAs quantum wires [5] prepared by cleaved edge overgrowth technique. Also, a Luttinger-liquid behavior has been reported in tunneling into InSb wires naturally grown in a porous material (asbestos) [6]. Tunneling pseudogap of a quantum wire has been described in terms of a Luttinger-liquid model both for a single- [7] and multi-subband wire [8], the latter system exhibiting a smooth healing of the pseudogap as the number of the occupied channels increases. In anticipation of more and better controlled tunneling experiments on quantum wires, and also from a general point of view, we would like to ask if there are any processes which could open a true gap, but not a pseudogap, in the electron spectrum of a wire, and if yes, what are the properties of the corresponding gapped phases.

To this end, we consider in this paper a two-subband quantum wire, having in mind semiconductor nanostructures studied recently in, e.g., Refs. [9–11]. To some approximation, this system is similar to other two important classes of 1D two-band systems studied extensively over the last few years, i.e., two-leg Hubbard ladders [12–18] and (single-wall) carbon nanotubes [19–23] (for an account of earlier results on coupled 1D systems, see, e.g., [24]). Studies of two-leg Hubbard ladders provided the classification of scattering processes in 1D two-subband systems and identification of processes capable of driving the system into a gapped state. Phase diagrams of a generic ladder, containing a multitude of gapped states, were constructed in Refs. [14–16]. A similarly formulated problem, with applications to a 1D system with electron and hole bands (valence-fluctuation problem), was investigated some time ago in [25].

The goal of the present paper is two-fold. First of all, we would like to understand which of the gapped phases, found in Hubbard ladders, have a chance to occur in quantum wires. The main difference between these two systems is that the Coulomb interaction in wires is (i) (supposed to be) purely repulsive; (ii) relatively long-ranged (even in the presence of a metallic gate); (iii) relatively well-known at distances larger than the lattice spacing (which is the range relevant for quantum wires); this imposes constraints on the choice of coupling constants for the Hubbard model. Also, because the electron wavelength in wires is larger than the lattice spacing, Umklapp scattering is unimportant. All these constraints reduce the variety of possible gapped states to (a) inter-subband charge-density wave (CDW), and (b) superconducting state. [We will come back to a more detailed description of these states shortly.] The main difference between carbon nanotubes and quantum wires is that the former, because of its special crystal structure, has two conducting subbands with *commensurate* Fermi-momenta. Although, as we will show, a quantum wire with nominally different subband Fermi-momenta may be driven into the commensurate state by inter-subband backscattering, this process occurs in a competition with other processes and requires special analysis.

Having determined which gapped phase can in principle occur in a quantum wire, we focus on the calculation of measurable quantities in each of these phases, which is the second goal of the present paper. Our main emphasis is on the tunneling density of states, which turns out to exhibit an unusual threshold behavior due to coexistence of gapped and gapless modes and also be sensitive to the presence of open boundaries. In addition, we consider the two-terminal conductance both in the absence and in the presence of impurities.

Although this paper is not supposed to be a review, we present, when possible, derivations of important results by other authors, e.g., Refs. [22,26–28] adopted to the context of our study. Hopefully, this would help a reader, who is not an expert in this field, to understand connections between different approaches.

Having formulated the goals and scope of this paper, we now return to a generic two-subband quantum wire with *incommensurate* Fermi-momenta k_{1F} and k_{2F} in subbands 1 and 2, respectively. In the basis of occupied transverse states, this becomes the problem of two Luttinger liquids coupled by inter-subband interactions. To understand possible phases of such a system, one should consider all possible scattering processes involving electrons from different subbands, i.e., forward scattering, backscattering, and "Cooper scattering"(cf. Figures 1,2,3). Forward scattering simply renormalizes the parameters of Luttinger liquids formed by electrons of each subbands but does not result in new phases, although it does change the conditions for occurrence of new phases. (To be more precise, forward scattering between Luttinger liquids in different subbands is responsible for the crossover into the Fermi-liquid state, but this crossover occurs smoothly as the number of channels increases).

The momentum transfer in a backscattering event involving electrons of different subbands is equal to $k_{1F} \pm k_{2F}$. If $|k_{1F} - k_{2F}| \gg T/\min\{v_{1F}, v_{2F}\}$, then there are no final states available for electrons involved in such a process (here v_{1F} and v_{2F} are the Fermi velocities in the two subbands). Thus, if the temperature T is low enough, interchannel backscattering is forbidden. However, it may become energetically favorable for a system to equalize the charge densities, and hence the Fermi momenta of different subbands. This may occur if the Fermi-momenta difference is small enough and the amplitude of backscattering is large enough. After the densities are adjusted, backscattering becomes possible. As a result, inter-subband charge-density wave (CDW) phase may be formed, in which charge densities of the subbands form a staggered pattern, see Fig. 4. Similarly to classical charge-density waves, this phase is very sensitive to a random potential, resulting in pinning of the CDW and strong suppression of conductance with disorder.

The "Cooper" scattering event, on the other hand, always conserves momentum and energy. In this process, two electrons starting in, *e.g.*, subband 1 with momenta k_{1F} and $-k_{1F}$, scatter on each other and end up in the other subband, also with opposite momenta k_{2F} and $-k_{2F}$. "Cooper scattering" can be considered as formation of a fluctuational Cooper pair in one of the subband followed by its tunneling into the other one. When kinetic energy gain due to such tunneling overcomes Coulomb repulsion between electrons forming pairs, the wire is in a Cooper (or superconducting) phase. This phase is characterized by locking of fluctuating charge currents in different subbands to each other as well as by spin gaps in each of the subbands. The Cooper phase is favored when Fermi-momenta imbalance is largest, i.e. when the second subband just starts to fill up. Disorder has a less pronounced effect on Cooper phase than on CDW one, similarly to what happens in higher dimensions.

It is important to emphasize here that inter-subband backscattering and Cooper scattering block only modes describing relative charge- and spin- density and current excitations, but leave the center-of-mass charge mode free. As a result, conductance of a clean CDW and Cooper phases remains at the universal value of $2e^2/h$ per occupied subband.

Despite being ideal conductors, both the CDW and Cooper phases are characterized by the truly gapped behavior of the tunneling density of states at energies below corresponding gaps. This is so because a 1D electron decomposes into charge and spin collective density excitations, and if some of these elementary collective excitations acquire a gap, the entire electron acquires it as well.

Somewhat surprisingly, we find that tunneling into the end of the CDW wire is quite different. A tunnel barrier at the end of the wire distorts charge-density wave profile and creates a static semi-soliton. This allows tunneling into the end to occur even at energies below the bulk CDW gap.

In Section 6.2 we consider, for illustrative purposes, the density of states of the "Mott" phase, which occurs in a single- or multi-subband wire subject to an external periodic potential [29,30]. In the case of a semiconductor wire, this potential may be provided by an additional electrostatic gate of a periodic shape [31]. Varying the potential applied to this gate, one can tune electrons of the wire into the half-filling condition (one electron per unit cell of the periodic potential). Unlike the two strong-coupling phases mentioned above, the Mott phase, which is described by the half-filled Hubbard model, does not conduct current because its total charge fluctuations are gapped by the external potential.

2   Hamiltonian of a Two-Subband Quantum Wire

2.1   Classification of Scattering Processes

Electrons in a quantum wire are described by the following Hamiltonian

$$H = \sum_{s} \int d^3r \Psi_s^\dagger(\mathbf{r}) \left(-\frac{1}{2m}\nabla_r^2 - \mu + V_{conf}(\mathbf{r}_\perp) \right) \Psi_s(\mathbf{r})$$

$$+ \frac{1}{2}\sum_{s,s'} \int d^3r d^3r' U(\mathbf{r}-\mathbf{r}')\Psi_s^\dagger(\mathbf{r})\Psi_{s'}^\dagger(\mathbf{r}')\Psi_{s'}(\mathbf{r}')\Psi_s(\mathbf{r}), \qquad (1)$$

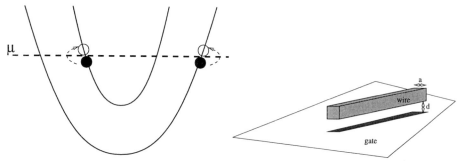

Fig. 1. Left: Example of inter-subband forward scattering. Filled (empty) circles denote initial (final) states of electrons. Dashed lines with arrows indicate "direction" of the scattering. Right: Schematic view of a gated wire. The wire of typical transverse size a is separated by distance d from the metallic gate. Distance d determines the range of interaction among electrons inside the wire.

where $V_{conf}(\mathbf{r}_\perp)$ is the confining potential in the transverse direction and $U(\mathbf{r})$ is the electron-electron interaction potential. The Fermi-wavelength of electrons in a semiconducting quantum wire is much larger than the lattice spacing of the underlying crystal structure. Because of that, we do not consider umklapp processes, in which electron momentum is transferred to the lattice (except for in Ch.6.2). Hamiltonian (1) is Galilean-invariant, and hence our subsequent calculations have to preserve this invariance as well. We will return to this important point later on in our discussion.

If the chemical potential in the leads is such that only two lowest subbands of transverse quantization are occupied, the electron wavefunction is given by

$$\Psi_s(\mathbf{r}) = \sum_{n=1}^{2} \phi_n(\mathbf{r}_\perp)\psi_{ns}(x), \tag{2}$$

where $\phi_n(\mathbf{r}_\perp)$ are the orthogonal wavefunctions of transverse quantization, which we choose to be real. In this basis, the kinetic part of Hamiltonian (1) becomes

$$H_0 = \sum_{n,s} \int dx\, \psi_{ns}^\dagger(x)\left(-\frac{\partial_x^2}{2m} - \mu + \epsilon_{ns}\right)\psi_n(x) \tag{3}$$

where ϵ_n is the energy of the n-th transverse subband.

To describe low-energy excitations in the n-th subband, we expand the longitudinal part of the Ψ-operator, $\psi_s(x)$, in terms of right- and left-moving excitations, residing around $\pm k_{nF}$ Fermi-points of the n-th channel:

$$\psi_{ns}(x) = R_{ns}(x)e^{ik_{nF}x} + L_{ns}(x)e^{-ik_{nF}x}. \tag{4}$$

In this representation, the interaction (four-fermion) part of Hamiltonian (1) reduces to a sum of two terms. The first one, U_{intra}, describes the interaction of electrons within the same subband, and contains usual forward and backward scattering processes. The second one, U_{inter}, describes the inter-subband inter-action. It splits naturally into *forward* (U^F), *backward* (U^B), and *Cooper* (U^C) parts.

$$U_{inter} = U^F + U^B + U^C. \tag{5}$$

Forward scattering involves no momentum transfer between subbands (cf. Fig.1). This process is also an example of a *direct* process, in a sense that electrons stay in the same subband, as is evident from the explicit expression for U^F

$$U^F = \frac{1}{2} \sum_{n \neq m} \int_{x,x'} M_d^{\{nm\}}(x - x') \sum_{s,s'} [R_{ns}^\dagger(x)R_{ns}(x) + L_{ns}^\dagger(x)L_{ns}(x)]$$
$$\times [R_{ms'}^\dagger(x')R_{ms'}(x') + L_{ms'}^\dagger(x')L_{ms'}(x')], \tag{6}$$

where the *direct* matrix element is given by

$$M_d^{\{nm\}}(x - x') = \int_{\mathbf{r}_\perp, \mathbf{r}'_\perp} U(\mathbf{r} - \mathbf{r}')\phi_n^2(\mathbf{r}_\perp)\phi_m^2(\mathbf{r}'_\perp); \tag{7}$$

and $\int_{z,z'} \equiv \int dz \int dz'$. By "backward scattering", we understand processes with a non-zero momentum transfer $\delta k = k_{1F} \pm k_{2F}$ between subbands. These processes can be divided further into *direct* and *exchange* parts

$$U^B = U_d^B + U_x^B, \tag{8}$$

In an *exchange* process [see Figs.2,3], electrons change subbands. Two parts of backscattering can be written as (for the sake of brevity, we omit here the $x - x'$-dependence of the matrix elements):

$$U_d^B = \frac{1}{2} \sum_{n \neq m} \sum_{s,s'} \int_{x,x'} M_d^{\{nm\}} [R_{ns}^\dagger(x)L_{ns}(x)L_{ms'}^\dagger(x')R_{ms'}(x')e^{2i(k_{mF}-k_{nF})(x+x')}$$
$$+ (R \Leftrightarrow L)e^{-2i(k_{mF}-k_{nF})(x+x')}] \tag{9}$$

and

$$U_x^B = -\frac{1}{2} \sum_{n \neq m} \sum_{s,s'} \int_{x,x'} M_x^{\{nm\}} [\{R_{ns}^\dagger(x)R_{ns'}(x')R_{ms'}^\dagger(x')R_{ms}(x)e^{i(k_{mF}-k_{nF})(x-x')}$$
$$+ (R \Leftrightarrow L)e^{-i(k_{mF}-k_{nF})(x-x')}\}$$
$$+ \{R_{ns}^\dagger(x)R_{ns'}(x')L_{ms'}^\dagger(x')L_{ms}(x)e^{i(k_{mF}+k_{nF})(x-x')}$$
$$+ (R \Leftrightarrow L)e^{-i(k_{mF}+k_{nF})(x-x')}\}], \tag{10}$$

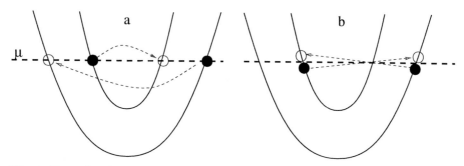

Fig. 2. Example of inter-subband backscattering: (a) direct, (b) exchange. Notations as in Fig.1.

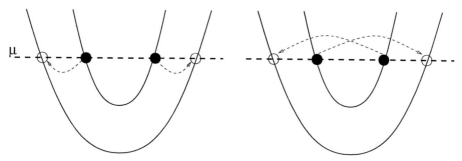

Fig. 3. Examples of inter-subband Cooper scattering. Notations as in Fig.1.

where the exchange matrix element is

$$M_x^{\{nm\}}(x-x') = \int_{\mathbf{r}_\perp, \mathbf{r}'_\perp} U(|\mathbf{r}-\mathbf{r}'|)\phi_n(\mathbf{r}_\perp)\phi_m(\mathbf{r}'_\perp)\phi_n(\mathbf{r}'_\perp)\phi_m(\mathbf{r}_\perp). \quad (11)$$

Momentum conservation requires that the energy of at least one of the states, involved into direct backscattering, should be far away from the Fermi energy, which forbids this process at not too high temperatures ($T \ll |k_{mF} - k_{nF}|\min\{v_{nF}, v_{mF}\}$). This is reflected in the presence of the exponential factors in front of the fermion operators in Eq.(9), which oscillate rapidly as functions of $(x+x')$. This restriction can be lifted though, if the system prefers to gain energy from backscattering by equalizing the subband densities, so that $k_{mF} = k_{nF}$.

Finally, we call "Cooper scattering"(Fig.3) a process in which two electrons with zero total momentum (a fluctuational Cooper pair) hop from, e.g., channel m, into channel n, so that the total momentum $Q = -k_m + k_m = 0 \to -k_n + k_n = 0$ is conserved. This process is also referred to as "Josephson coupling" [32,24], or as "$g_{00\pi\pi}$ process" [13],[16]. The Hamiltonian of Cooper scattering is given by

$$U^C = \frac{1}{2} \sum_{n \neq m} \sum_{s,s'} \int_{x,x'} M_x^{\{nm\}} [R_{ns}^\dagger(x) L_{ns'}^\dagger(x') e^{ik_{nF}(x'-x)} +$$
$$+ L_{ns}^\dagger(x) R_{ns'}^\dagger(x') e^{-ik_{nF}(x'-x)}] \times$$
$$\times [R_{ms'}(x') L_{ms}(x) e^{ik_{mF}(x'-x)} + L_{ms'}(x') R_{ms}(x) e^{-ik_{mF}(x'-x)}]. \quad (12)$$

By construction, Cooper scattering is of the *exchange* type.

In what follows, we use the following abbreviations: forward scattering $\equiv$ FS, direct backward scattering $\equiv$ dBS, exchange backward scattering $\equiv$ xBS, Cooper scattering $\equiv$ CS.

For a generic situation of $k_n \neq k_m$, the only momentum-conserving inter-subband scattering processes are FS, xBS, and CS. The amplitudes of these processes depend on the ratio a/d, where a is a typical transverse size of the wire (which determines the spatial extension of $\phi_n(\mathbf{r}_\perp)$) and d is the range of interactions. In the limit $a/d \to 0$, the interaction potential can be taken out of integrals (7) and (11), upon which M_d remains finite, whereas M_x vanishes. It can be readily shown that for finite but small ratio a/d the exchange matrix element is small: $M_x \sim (a/d)^2 M_d$. The long-range interaction thus discriminates against exchange processes. If (as it is most often the case) a wire is formed by means of a gate deposited over the 2D heterostructure, d is given by the distance to this gate, which screens the Coulomb interaction in the wire (see Fig. 1). Typically, $a/d = 0.1 - 1$.

2.2   Bosonized Form of the Hamiltonian

We use the conventional bosonization procedure, in which

$$R_{ns}(x) = \frac{1}{\sqrt{2\pi\alpha}} e^{i\sqrt{\pi}(\varphi_{ns} - \theta_{ns})}, \quad (13)$$

$$L_{ns}(x) = \frac{1}{\sqrt{2\pi\alpha}} e^{-i\sqrt{\pi}(\varphi_{ns} + \theta_{ns})}, \quad (14)$$

and short-range cut-off $\alpha \sim k_F^{-1}$. Boson fields φ_{ns} and θ_{ns} with $n = 1, 2$; $s = \pm 1$, are decomposed into charge-(ρ) and spin-(σ) collective modes

$$\varphi_{ns} = \frac{1}{\sqrt{2}}(\varphi_{n\rho} + s\varphi_{n\sigma}),$$

$$\theta_{ns} = \frac{1}{\sqrt{2}}(\theta_{n\rho} + s\theta_{n\sigma}). \quad (15)$$

Parts of the Hamiltonian, describing the free motion and intra-subband interactions ($H_0 + U_{intra}$), take the well-known Luttinger-liquid form:

$$H_{n\rho} = \frac{1}{2}\int dx \{v_{n\rho}K_{n\rho}(\partial_x\theta_{n\rho})^2 + \frac{v_{n\rho}}{K_{n\rho}}(\partial_x\varphi_{n\rho})^2\}, \tag{16}$$

$$H_{n\sigma} = \frac{1}{2}\int dx \{v_{n\sigma}K_{n\sigma}((\partial_x\theta_{n\sigma})^2 + \frac{v_{n\sigma}}{K_{n\sigma}}(\partial_x\varphi_{n\sigma})^2\}$$
$$+\frac{2U(2k_{nF})}{(2\pi\alpha)^2}\int dx \cos[\sqrt{8\pi}\varphi_{n\sigma}], \tag{17}$$

which describes *independent* charge- and spin-density excitations ($H_{n\rho}$ and $H_{n\sigma}$, respectively). The cosine term in Eq.(17) is due to backscattering within a single subband. Explicit expressions for the Luttinger-liquid parameters will be discussed later.

Upon bosonization, the three types of intersubband interactions take the following form :

$$U^F = \frac{2f_0}{\pi}\int dx \partial_x\varphi_{1\rho}\partial_x\varphi_{2\rho}; \tag{18}$$

$$U_d^B = \frac{4f_{bs}}{\pi^2\alpha^2}\int dx \cos\left[\sqrt{2\pi}(\varphi_{1\rho}-\varphi_{2\rho}) + 2(k_{1F}-k_{2F})x\right]$$
$$\times \cos[\sqrt{2\pi}\varphi_{1\sigma}]\cos[\sqrt{2\pi}\varphi_{2\sigma}]; \tag{19}$$

$$U_x^B = -\frac{1}{2}\int dx \left(\frac{b_1+b_2}{\pi}(\partial_x\varphi_{1\rho}\partial_x\varphi_{2\rho} + \partial_x\varphi_{1\sigma}\partial_x\varphi_{2\sigma})\right.$$
$$+ \frac{b_1-b_2}{\pi}(\partial_x\theta_{1\rho}\partial_x\theta_{2\rho} + \partial_x\theta_{1\sigma}\partial_x\theta_{2\sigma})$$
$$+\frac{2}{\pi^2\alpha^2}\cos[\sqrt{2\pi}(\theta_{1\sigma}-\theta_{2\sigma})]\{(b_1+b_2)\cos[\sqrt{2\pi}\varphi_{1\sigma}]\cos[\sqrt{2\pi}\varphi_{2\sigma}] -$$
$$\left.(b_1-b_2)\sin[\sqrt{2\pi}\varphi_{1\sigma}]\sin[\sqrt{2\pi}\varphi_{2\sigma}]\}\right); \tag{20}$$

$$U^C = \frac{4}{2\pi^2\alpha^2}\int dx \{t_{sp}\cos[\sqrt{2\pi}(\theta_{1\rho}-\theta_{2\rho})]\cos[\sqrt{2\pi}\varphi_{1\sigma}]\cos[\sqrt{2\pi}\varphi_{2\sigma}] + \tag{21}$$
$$+ t_{tp}\cos[\sqrt{2\pi}(\theta_{1\rho}-\theta_{2\rho})]\left(\cos[\sqrt{2\pi}(\theta_{1\sigma}-\theta_{2\sigma})] - \sin[\sqrt{2\pi}\varphi_{1\sigma}]\sin[\sqrt{2\pi}\varphi_{2\sigma}]\right)\}.$$

The corresponding amplitudes are given by

$$f_0 = \int dx M_d^{\{12\}}(x),$$
$$f_{bs} = \int dx M_d^{\{12\}}(x)\cos[(k_{1F}+k_{2F})x],$$
$$b_{1,2} = \int dx M_x^{\{12\}}(x)\cos[(k_{1F}\mp k_{2F})x],$$
$$t_{sp} = \int dx M_x^{\{12\}}(x)\cos(k_{1F}x)\cos(k_{2F}x),$$
$$t_{tp} = \int dx M_x^{\{12\}}(x)\sin(k_{1F}x)\sin(k_{2F}x). \tag{22}$$

In the last two lines, $t_{sp}(t_{tp})$ are the amplitudes of singlet (triplet) Cooper processes.

The highly non-linear (cosine) terms in Eqs.(19,20,21) signal potential instabilities of the ground state due to interactions. For the dBS process [Eq.(19)], this instability is of the charge-density-wave (CDW) type, quantity $\varphi_{1\rho} - \varphi_{2\rho}$ being the phase of the CDW (particle-hole) condensate. If subbands are equivalent ($k_{1F} = k_{2F}$), the energy is minimized by adjusting the CDW-condensate phase is such a way that the cosine takes its minimum value (-1, for repulsive interactions). For non-equivalent subbands ($k_{1F} \neq k_{2F}$), the global minimization of the energy is impossible due to the position-dependent phase shift, and thus the CDW instability is suppressed. Nevertheless, if the energy gain due to opening of the CDW gap is large enough, the system may choose to adjust the subband densities, which makes the CDW instability possible. Density equilibration is most likely to occur if the cross-section of the wire is approximately symmetric. For example, if it is a perfect square, the ground state is doubly degenerate. Deformation lifts the degeneracy but the energy splitting between the states is small for small deformations. Such states are almost equally occupied, and a small difference in densities is likely to be eliminated by opening the CDW-gap. It seems that cleaved edge quantum wires investigated by Yacoby et al. [11] satisfy this requirement.

In the "two-chain model", the CDW-process of this type is known as "deconfinement"[13]: degeneracy of bonding and antibonding subbands implies that the amplitude of interchain tunneling, $t_\perp$, is renormalized to zero by interactions, and electrons thus remain "confined" to their respective chains.

The xBS process (20) contains both harmonic terms, arising from backscattering of electrons with parallel spins, and cosine terms, arising from backscattering with antiparallel spins. The latter contain only spin fields and thus can lead to the instability only in the spin channel. In the terminology of Ref. [16,14], this instability corresponds to the "orbital antiferromagnet phase"(OAF). The OAF instability occurs only if backscattering is sufficiently strong [16,14]. For a quantum wire, in which all amplitudes are given just by the corresponding Fourier components of the same interaction potential, this conditions means that $U(2k_F) > 2U(0)$ (for identical subbands), which is never the case for any physical $U(\mathbf{r})$. In what follows, we will not therefore consider the OAF phase. (Note that in a "two-chain model", amplitudes of various scattering processes may be determined by entirely different physics, e.g., some of them may result from direct electron-electron interaction and some from exchange of virtual phonons. Hence, the ratios of amplitudes may be arbitrary, and the OAF instability is possible, at least *a priori*.)

Finally, the CS process [Eq.(21)] may lead to a superconducting instability (of both singlet and triplet types), accompanied by opening of spin gaps in each of the subbands, in a analogy with a superconducting transition in higher dimensions. The quantity $\theta_{1\rho} - \theta_{2\rho}$ plays the role of the superconducting condensate phase. Inter-subband forward scattering [Eq. (18)] plays an important role in developing a superconducting instability–it reduces electron repulsion in the relative charge-density fluctuation channel, making it possible for Cooper scattering to become relevant. The superconducting phase is also known as "C1S0-phase"(meaning: one gapless charge mode and no gapless spin modes) [15] or "d-wave superconductor"(indicating that order parameter is odd upon interchanging subband index of electrons, forming a Cooper pair) [16,14]. This particular instability received much attention recently as one of the models of HTC superconductivity [17]. We also note in passing that the idea of super-conductivity in a two-band system has a long history, starting from the 1968 paper by Frölich [33] (for a review, see Ref. [34]). The idea, employed in earlier work, is that if the masses of electrons in two subbands are significantly different, there always–even in 3D–exists a gapless plasmon excitation (a direct analog of a Langmuir-Tonks ion sound wave in plasma), which serves as a mediator of effective attraction. The superconducting phase in 1D two-band system is already composed of gapless excitations and is not limited by the condition of different masses (although, as we will see shortly, there is no lack of other constraints).

Note also that U_d^B (19) and U^C (21) mix charge and spin modes, and thus spoil the spin-charge separation present in the Hamiltonian of a single subband.

Processes (18 - 21) have been written down in the literature in many different ways, so it is worth to make a connection to previous work here. Identification of our notations with the g-ology ones (used by Schulz in his two important papers [16,14]) is as follows: $t_{sp} = g_{12} + g_{23}$, $t_{tp} = g_{23} - g_{12}$, $b_1 = g_{13}$, and his g_{11}-process (backscattering with opposite spins) should be equated with our $U(2k_{nF})$ in Eq.(17). There is no correspondence to our amplitude b_2, which describes exchange inter-subband backscattering of the type $R_{ns}^\dagger R_{ms'}^\dagger R_{ns'} R_{ms} + (R \to L)$, see Eq.(10). That such a process is absent in Ref. [16] is clear from Eq.(2) of that reference.

We can make one more connection by noting that (somewhat lengthy) Eq.(20) can be also represented compactly as

$$U_x^B = - \int dx \{(b_1 + b_2)(\rho_1 \rho_2 + S_1 S_2) + (b_1 - b_2)(j_{c1} j_{c2} + j_{s1} j_{s2})\}, \qquad (23)$$

where ρ_n (S_n) is the charge (spin) density, and j_{cn} (j_{sn}) is the charge (spin) current in the n-th subband, using notations of Emery, Kivelson, and Zachar [17].

3   Spinless Electrons

Model In this section we consider a "toy"model of two subbands of spinless electrons, which contains all interesting effects we want to discuss and, at the same time, allows a rather complete analytic treatment. In this model, various parts of the Hamiltonian reduce to

$$H_0 \rightarrow \tilde{H}_0 = \int dx \sum_n \left[\frac{v_n}{2K_n}(\partial_x \varphi_n)^2 + \frac{v_n K_n}{2}(\partial_x \theta_n)^2 \right] \tag{24}$$

$$U^F \rightarrow \tilde{U}^F = \frac{f_0}{\pi} \int dx \partial_x \varphi_1 \partial_x \varphi_2 \tag{25}$$

$$U_d^B \rightarrow \tilde{U}_d^B = \frac{f_{bs}}{\pi^2 \alpha^2} \int dx \cos[\sqrt{4\pi}(\varphi_1 - \varphi_2) + 2(k_{1F} - k_{2F})x] \tag{26}$$

$$U_x^B \rightarrow \tilde{U}_x^B = -\frac{1}{2\pi} \int dx \left[(b_1 + b_2)\partial_x \varphi_1 \partial_x \varphi_2 + (b_1 - b_2)\partial_x \theta_1 \partial_x \theta_2 \right] \tag{27}$$

$$U^C \rightarrow \tilde{U}^C = \frac{f_C}{2\pi^2 \alpha^2} \int dx \cos \sqrt{4\pi}(\theta_1 - \theta_2); \tag{28}$$

so that

$$\tilde{H} = \tilde{H}_0 + \tilde{U}^F + \tilde{U}_d^B + \tilde{U}_x^B + \tilde{U}^C. \tag{29}$$

Amplitude f_C plays now the role of t_{sp} for spinless electrons. Analysis of potentially "dangerous"(in a sense of inducing instabilities) intersubband processes reduces to estimating the scaling dimensions of corresponding cosine operators in terms of the parameters of the harmonic part. In their turn, these parameters are related to the Fourier components of the electron-electron interaction potential. As it turns out, the latter relation is not that straightforward, and we will clarify this point in the next Section.

Galilean invariance and Pauli principle: single-subband Luttinger liquid To begin with, we consider the simplest case when there is no intersubband interaction and the Hamiltonian is given by the sum of two single-subband Hamiltonians (24). (As our discussion is referred now to a single subband, we suppress temporarily the subband index.) For a given effective 1D interaction potential $U(x)$, the Luttinger-liquid parameters (K and v) depend on the $q = 0$ and $q = 2k_F$ Fourier components of U, as well on the bare Fermi velocity v_F:

$$K = \mathcal{K}[U(0)/v_F, U(2k_F)/v_F], \tag{30}$$
$$v = v_F \mathcal{V}[U(0)/v_F, U(2k_F)/v_F], \tag{31}$$

where $\mathcal{K}(x, y)$ and $\mathcal{V}(x, y)$ are some dimensionless functions of their arguments. Relations (30),(31) have to satisfy (i) the Pauli principle and (ii) Galilean invariance. The Pauli principle for spinless fermions means that for the case of contact interaction, i.e., when $U(0) = U(2k_F)$, the system should behave as if there is no interaction at all. Accordingly, $K = 1$ and $v = v_F$ for this case, or

$$\mathcal{K}(x, x) = \mathcal{V}(x, x) = 1. \tag{32}$$

Galilean invariance stipulates that $Kv = v_F$, or

$$\mathcal{K}(x, y)\mathcal{V}(x, y) = 1, \forall x, y. \tag{33}$$

Physically, condition (33) comes about either by requiring that the shift of the ground state energy due to the motion of a system as a whole does not depend on the interaction [35], or by requesting that the dc conductivity of a uniform system[0] does not depend on interactions (Peierls theorem) [1,36]. (Also, one can use the interaction-invariance of the persistent current in a ring threaded by the Aharonov-Bohm flux).

Conventional bosonization of the g-ology Hamiltonian (see, e.g., review [1]) leads to

$$K = \sqrt{\frac{2\pi v_F + g_4 - g_2}{2\pi v_F + g_4 + g_2}}, \quad v = v_F \sqrt{(1 + \frac{g_4 - g_2}{2\pi v_F})(1 + \frac{g_4 + g_2}{2\pi v_F})}. \quad (34)$$

In terms of the Fourier components of the interaction potential the g-parameters are expressed as $g_4 = U(0)$ (right-right and left-left amplitude) and $g_2 = U(0) - U(2k_F)$ (right-left amplitude), and Eq.(34) gives

$$K = \left[\frac{1 + \frac{U(2k_F)}{2\pi v_F}}{1 + \frac{2U(0) - U(2k_F)}{2\pi v_F}}\right]^{1/2},$$

$$v = v_F \left[1 + \frac{U(2k_F)}{2\pi v_F}\right]^{1/2} \cdot \left[1 + \frac{2U(0) - U(2k_F)}{2\pi v_F}\right]^{1/2}. \quad (35)$$

One can see that the expressions above do not satisfy conditions (32), (33). Indeed, it follows from Eq. (35) that $v \neq v_F$ for $U(0) = U(2k_F)$ and that $Kv \neq v_F$ as long as $U(2k_F) \neq 0$. Usually, the spinless Luttinger liquid model does not include backscattering explicitly. The rationale for such a simplification is that for spinless particles in 1D this process is indistinguishable from forward scattering, see, e.g., Ref. [37]. We do not find this approach satisfactory, as it is clear that the behavior of the system should be determined both by forward and backward amplitudes. Also, correct expressions for K and v should include both $U(0)$ and $U(2k_F)$, otherwise the Pauli principle cannot be satisfied. This argument can also be re-phrased in terms of direct and exchange contributions to the self-energy [38].

What did we do wrong to arrive at the Luttinger-liquid model which does not satisfy two basic physical principles? As one can show by using the Ward indentities (conservation laws) for the system of interacting electrons with *linear* spectrum [37], the problem occurs already at the level of fermions and is thus not inflicted by some subtleties of bosonization. Rather, it is a manifestation of an *anomaly*, i.e., a violation of the conservation law caused by regularization, which one is forced to used in a model with linear and unbound spectrum [37].

One way to deal with this problem is to replace Eqs.(35) by expressions which do not follow directly from the original fermion Hamiltonian with linear spectrum, but do satisfy all necessary criteria. This is an accord with the point of view [40] that one should consider K and v as phenomenological parameters,

---

[0] Here we consider a uniform Luttinger liquid. The role of reservoirs, to which the wire is attached to, will be discussed in Sec.5.

which are renormalized from their bare values by irrelevant or marginal operators neglected in the course of linearization. We will be able to find *exact* expressions for K and v, satisfying the minimal set of requirements.

First, we notice that the Pauli principle requires $\mathcal{K}(x, y)$ to be a function of either $x - y$ or x/y. The latter choice contradicts to the requirement that $\mathcal{K}$ must have Taylor expansions both around $x = 0$ and $y = 0$. Therefore,

$$\mathcal{K}(x, y) = \kappa(x - y), \tag{36}$$
$$\kappa(0) \quad = 1. \tag{37}$$

Then we notice that the model with forward scattering only, i.e., the original Luttinger model [39], does respect Galilean invariance. Therefore one can take the Luttinger model expression for K as the correct one, which means that

$$\mathcal{K}(x, 0) = 1/\sqrt{1 + x}. \tag{38}$$

Combining Eq. (37) with Eq. (38), we see that

$$\mathcal{K}(x, y) = 1/\sqrt{1 + x - y}, \tag{39}$$
$$\mathcal{V}(x, y) = \sqrt{1 + x - y}, \tag{40}$$

or,

$$K = \left[1 + \frac{U(0) - U(2k_F)}{\pi v_F}\right]^{-1/2}, \tag{41}$$

$$v = v_F \left[1 + \frac{U(0) - U(2k_F)}{\pi v_F}\right]^{1/2}. \tag{42}$$

The physical meaning of Eq.(41),(42) is obvious: the effective interaction is equal to backscattering minus forward scattering. One can check that a Luttinger-liquid model with parameters given by Eq.(41),(42) reproduces correctly results for a 1D electron system, obtained without linearization but in the limit of weak interactions. For instance, the (inverse) compressibility of a Luttinger liquid, parametrized by K and v from Eq.(41),(42), is given by

$$\frac{1}{\chi} = \frac{\pi v}{K} = \pi v_F + U(0) - U(2k_F). \tag{43}$$

As one can check, Eq.(43) coincides with the inverse compressibility of electrons with a *quadratic* spectrum obtained in the Hartree-Fock approximation. A perturbative (linear in U) form of Eq.(42) has recently been derived in [41]. It can also be read of from the tunneling exponent of a 1D system with a *quadratic* dispersion [42].

Galilean invariance and Pauli principle: coupled subbands Now we allow for harmonic coupling between subbands, i.e., take into account intersubband forward [Eq.(25)] and exchange backscattering [Eq.(27)]. For the contact inter-action, the amplitudes of these two processes coincide: $b_1 = b_2 = f_0$ and, as a result, inter-subband interaction drops out. The Pauli principle is thus satisfied. Intersubband exchange backscattering does violate Galilean invariance, and the correction procedure, similar to that for a single subband, is necessary. We will not do it here however, because for a long-range interaction $(a/d \ll 1)$, the vio-lation is "weak": the deviation from the Galilean-invariant result is proportional to the *exchange* amplitudes, which are small compared to the *direct* ones.

3.1   Nearly Equivalent Subbands

First, we discuss the CDW-instability, which may occur if the density equilibra-tion between subbands is energetically favorable. To simplify the discussion, we consider the case of a long range interaction $(a/d \ll 1)$, when amplitudes of all exchange processes are small. In the leading order, $f_C = b_{1,2} = 0$ and the only "dangerous" process to be considered is direct intersubband backscattering. Furthermore, we assume that subbands are nearly equivalent and put $v_{1F} = v_{2F}$ and $K_1 = K_2$ but keep $\delta k_F = k_{1F} - k_{2F}$ in Eq. (26) finite. It is convenient to introduce symmetric and antisymmetric combinations of boson fields

$$\varphi_{\pm} = \frac{\varphi_1 \pm \varphi_2}{\sqrt{2}}; \theta_{\pm} = \frac{\theta_1 \pm \theta_2}{\sqrt{2}}, \qquad (44)$$

which correspond to fluctuations of total (+) and relative (-) subband charge and current. In terms of these fields,

$$
\begin{aligned}
H &= H_+ + H_-, \\
H_+ &= \frac{1}{2} \int dx \left\{ \frac{v_+}{K_+} (\partial_x \varphi_+)^2 + v_+ K_+ (\partial_x \theta_+)^2 \right\}, \qquad (45) \\
H_- &= \frac{1}{2} \int dx \left\{ \frac{v_-}{K_-} (\partial_x \varphi_-)^2 + v_- K_- (\partial_x \theta_-)^2 + \frac{f_{bs}}{\pi^2 \alpha^2} \cos[\sqrt{8\pi} \varphi_- + 2\delta k_F x] \right\} (46)
\end{aligned}
$$

where

$$K_+ = \left[1 + \frac{2U(0) - U(2k_F)}{\pi v_F} \right]^{-1/2} \quad \text{and} \quad K_- = \left[1 - \frac{U(2k_F)}{\pi v_F} \right]^{-1/2}, \qquad (47)$$

and $v_{\pm} = v_F / K_{\pm}$. Note that $K_- > 1$ for $U(2k_F) > 0$, which signals effective attraction in the $(-)$ channel.

Collective adjustment of densities as a commensurate-incommensurate transition To understand how the CDW-instability works, we consider first a model situation when $K_- < 1$, so that operator $\cos[\sqrt{8\pi}\phi_-]$ is relevant in the RG sense. Finite δk_F stops the RG-flow at scale $\ell \sim 1/\ln|\delta k_F|\alpha$, thus precluding the system from reaching its strong-coupling limit. However, this consideration

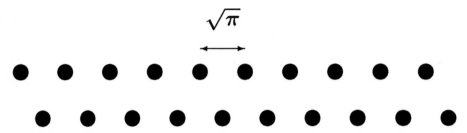

Fig. 4. An illustration of the charge density wave in two coupled subbands. A staggered configuration lowers the energy due to short-range repulsion, if the densities are commensurate.

does not take into account the possibility of a collective density readjustment between subbands. Such a readjustment may occur, if the kinetic energy loss $\Omega = v_F \delta k_F$ is compensated by the gain in the potential energy due to opening of the gap in the (-) channel. In other words, when the difference in electron densities is sufficiently small, the total energy is minimized by equating the densities and opening the charge gap. Obviously, such process cannot be considered at the level of single-particle description of transverse quantization. Instead, one should now treat eigenstates and eigenenergies of the wire as being determined by a self-consistent procedure, involving both single-particle and many-body effects.

The mechanism described above can be considered as a commensurate-incommensurate transition. The incommensurability, defined as $\mathcal{I} = L^{-1}\langle \int dx \partial_x \varphi_- \rangle$, where L is the length of the wire, is known to have a threshold behavior [43,44]: $\mathcal{I} \sim \sqrt{\Omega^2 - \Omega_c^2}\,\Theta(\Omega - \Omega_c)$, where $\Omega_c = \sqrt{2\pi K_-}\Delta_{CDW}$ and the gap follows from mapping on exactly solvable Heisenberg spin chain [45], $\Delta_{CDW} \sim (f_{bs})^{1/2(1-K_-)}$. As follows from the definition of the incommensurability, $\mathcal{I} = 0$ implies $\delta k_F = 0$, i.e., equal subband densities. Therefore, the re-adjustment takes place if $\Omega < \Omega_c$. Backscattering is then enabled and relevant (for $K_- < 1$), even for a non-zero initial value of δk_F.

What is the physical meaning of this instability? A simple picture can be obtained in the limit of strong (both inter- and intra-subband) interactions, when the potential energy dominates over quantum fluctuations. In this case, electrons of each of the subbands form a regular lattice (Wigner crystal). Boson fields φ_n also have periodic structures with period equal to $\sqrt{\pi}$ (recall that a shift of $\sqrt{\pi}$ corresponds to adding one electron into the system). For $f_{bs} > 0$, the energy of intrasubband repulsion

$$f_{bs} \cos[\sqrt{8\pi}\varphi_-] = -f_{bs} \cos[\sqrt{4\pi}(\varphi_1 - \varphi_2 + \sqrt{\pi}/2)] \qquad (48)$$

is minimized by a relative phase shift of $\sqrt{\pi}/2$ between the subbands, which corresponds to a shift of electron lattices by half-a-period. This is an intersubband charge-density wave (CDW).

Competition between CDW and Cooper channels Let us now suppose that the density re-adjustment did occur, i.e., $\delta k_F = 0$, but Cooper scattering is also present, so that the Hamiltonian of the (-)-channel is

$$H_- = \frac{1}{2} \int dx \left\{ \frac{v_-}{K_-}(\partial_x\varphi_-)^2 + v_- K_-(\partial_x\theta_-)^2 \right.$$
$$\left. + \frac{f_{bs}}{\pi^2\alpha^2}\cos\sqrt{8\pi}\varphi_- + \frac{f_C}{2\pi^2\alpha^2}\cos\sqrt{8\pi}\theta_- \right\}. \tag{49}$$

Which of the two instabilities–CDW or superconductivity–wins? The situation of this type, when cosines of both mutually conjugated fields (φ_- and θ_-) are present, was analyzed by Schulz and Giamarchi [46]. They found that the result is very sensitive not only to the value of K_-, which determines the scaling dimensions of the fields, but also to the ratio of amplitudes, f_C/f_{bs}. As $K_- > 1$ for repulsive $U(\mathbf{r})$, it may seem that superconductivity is favored over CDW. The situation is not that straightforward, however. For example, consider the situation of weak and long-range interactions, i.e., assume that $U(0), U(2k_F) \ll v_F$ and $a \ll d$. Because the interaction is weak, both processes are almost marginal, CDW being on the slightly irrelevant and superconductivity on the slightly relevant side. For long-range interactions, $f_{bs} \sim U(2k_F)$ and $f_C \sim (a/d)^2 U(0)$. Modeling $U(x)$ by

$$U(x) = \begin{cases} e^2/\epsilon x, & \text{for } x < d; \\ 0, & \text{for } x > d, \end{cases}$$

we get $U(2k_F)/U(0) \sim \ln(k_F a)/\ln(d/a)$. Thus

$$\frac{f_C}{f_{bs}} \sim \left(\frac{a}{d}\right)^2 \frac{\ln d/a}{\ln k_F a} \ll 1. \tag{50}$$

The RG-equation for K_- [46]

$$\frac{d}{dl}K_- = f_C^2 - f_{bs}^2 \tag{51}$$

shows that K_- decreases, if $|f_C| < |f_{bs}|$. Even if initially $K_-(0) > 1$, the situation with $K_-(l) < 1$, when CDW is relevant, will be reached in the process of renormalization. For weak and long-ranged interactions, CDW thus wins over superconductivity.

If interactions are not sufficiently weak and/or long-ranged, only a full RG solution can determine the leading instability. We will not analyze the general case here.

3.2   Non-Equivalent Subbands: Renormalization Group

Now we consider a generic situation of non-equivalent subbands, when δk_F is not small enough for the density re-adjustment to occur. We find that a strong imbalance between Fermi-velocities of occupied subbands actually helps super-conducting instability to develop (see *case B* below), despite the fact that a naive scaling dimension estimate does not show this. This effect follows from the next-to-leading order perturbative RG calculations, which we present here.

Because $\delta k_F \neq 0$, we neglect the dBS process [Eq. (26)] from the outset but keep the Cooper one [Eq. (28)]. For long-range interaction, one can also neglect the xBS process, Eq.(27), whose amplitude is small for this case: $b_i \propto (a/d)^2$. Its inclusion is straightforward (U_x^B is quadratic) but does not lead to any qualitatively new results, while complicating the analysis significantly. The Hamiltonian then reads

$$\tilde{H} = \tilde{H}_0 + \tilde{U}^F + \tilde{U}^C. \tag{52}$$

Because $\tilde{U}^C$ contains θ-fields, it is convenient to switch from the Hamiltonian to the Lagrangian approach and to integrate out the φ-fields. The quadratic part of the resulting action is diagonalized by the following transformation

$$\begin{pmatrix} \bar{\theta}_1 \\ \bar{\theta}_2 \end{pmatrix} = \begin{pmatrix} \mu_1 & 0 \\ 0 & \mu_2 \end{pmatrix} \begin{pmatrix} \cos\beta & \sin\beta \\ -\sin\beta & \cos\beta \end{pmatrix} \begin{pmatrix} \sqrt{v_1 K_1}\theta_1 \\ \sqrt{v_2 K_2}\theta_2 \end{pmatrix}, \tag{53}$$

where

$$\tan 2\beta = \frac{u_0^2}{v_1^2 - v_2^2}, \quad u_0 = \left[2f_0\sqrt{v_1 K_1 v_2 K_2}/\pi\right]^{1/2}, \tag{54}$$

and

$$\mu_1 = \frac{\cos\beta}{\sqrt{v_1 K_1}} - \frac{\sin\beta}{\sqrt{v_2 K_2}},$$
$$\mu_2 = \frac{\sin\beta}{\sqrt{v_1 K_1}} + \frac{\cos\beta}{\sqrt{v_2 K_2}}. \tag{55}$$

In terms of new fields, the action is given by

$$S = \frac{1}{2}\int dx d\tau \left[\sum_n R_n\{\frac{1}{u_n}(\partial_\tau \bar{\theta}_n)^2 + u_n(\partial_x \bar{\theta}_n)^2\} + \frac{f_C}{\pi^2 \alpha^2}\cos\sqrt{4\pi}(\bar{\theta}_1 - \bar{\theta}_2)\right], \tag{56}$$

where

$$u_n^2 = \frac{1}{2}\left(v_1^2 + v_2^2 \pm \sqrt{(v_1^2 - v_2^2)^2 + u_0^4}\right) \tag{57}$$

are the velocities of new collective modes and $R_n = 1/(u_n\mu_n^2)$ are the new stiffness coefficients.

We are now ready to perform the momentum-shell RG, i.e., to develop perturbatively in coupling constant f_C and integrate out high-energy fluctuations with 2-momentum k within a thin strip $\Lambda - d\Lambda \leq k \leq \Lambda$ ($d\Lambda/\Lambda \ll 1$). The first-order contributions renormalize f_C, whereas the second-order one renormalize stiffnesses R_n. The main difference from the conventional RG-treatment of the sine-Gordon action (see, e.g., Ref. [47]) is that the f_C^2-contribution produces (among others) mixed gradient terms of the type $\partial_\nu \bar{\theta}_1 \partial_\nu \bar{\theta}_2$ ($\nu = \tau, x$), which are absent in bare action (56). To eliminate these terms, we transform fields one more time :

$$\begin{pmatrix} \tilde{\theta}_1 \\ \tilde{\theta}_2 \end{pmatrix} = \left(1 + \frac{\beta'}{t'}\right) \begin{pmatrix} \cosh \beta' & t' \sinh \beta' \\ \frac{1}{t'} \sinh \beta' & \cosh \beta' \end{pmatrix} \begin{pmatrix} \bar{\theta}_1 \\ \bar{\theta}_2 \end{pmatrix}, \tag{58}$$

where

$$t' = \frac{u_2 R_2}{u_1 R_1} \quad \text{and} \quad \beta' = \frac{d\Lambda}{\Lambda} \left(\frac{f_C}{\pi}\right)^2 \frac{1}{u_1 u_2 R_1 R_2} \tag{59}$$

are chosen in such a way that the coefficients in front of mixed gradient terms vanish. When written in terms of $\tilde{\theta}_n$, the action is brought into its original form but with renormalized parameters. The resulting RG equations read

$$\frac{d}{dl} \frac{1}{R_1} = -\bar{f}^2 \frac{1}{R_1} \left(\frac{4\gamma^2}{R_2^2(1+\gamma^2)} + \frac{2}{R_1^2(1+\gamma^2)} + \frac{\gamma}{R_1 R_2}\right), \tag{60}$$

$$\frac{d}{dl} \frac{1}{R_2} = -\bar{f}^2 \frac{1}{R_2} \left(\frac{4}{R_1^2(1+\gamma^2)} + \frac{2\gamma^2}{R_2^2(1+\gamma^2)} + \frac{1}{R_1 R_2 \gamma}\right), \tag{61}$$

$$\frac{d}{dl} \gamma = \bar{f}^2 \frac{1-\gamma^2}{R_1 R_2}, \tag{62}$$

$$\frac{d}{dl} \bar{f} = \left(2 - \frac{1}{R_1} - \frac{1}{R_2}\right) \bar{f}, \tag{63}$$

where $\bar{f} = (1/\pi) f_C \sqrt{u_1^{-2} + u_2^{-2}}$ is the dimensionless coupling constant and $\gamma = u_1/u_2$. To the $\bar{f}^2$-accuracy, all terms multiplying $\bar{f}^2$ on the right-hand-side of the first three equations above have to be treated as a constants determined by the initial conditions. The system of RG-equations has an obvious integral of motion

$$C = \frac{x^2}{c_1} + \frac{y^2}{c_2} - \bar{f}^2, \tag{64}$$

where $x = 1 - 1/R_1$, $y = 1 - 1/R_2$ and $c_{1,2}$ are the coefficients in front of $\bar{f}^2$ in Eqs. (60,61), respectively. Note also that $x = (c_1/c_2) y + p$ (we denote $x = x(l)$, $y = y(l)$ whereas initial conditions are denoted by sub-index 0, i.e., $x(0) = x_0$, etc.). Constants of motion C and p are determined by initial conditions.

The flow described by (60-63) is quite similar to that of a canonical Kosterlitz-Thouless system: x and y increase with $\bar{f}$ regardless of its sign. If x_0, $y_0 > 0$, $\bar{f}$ grows unrestrictedly, flowing into the strong-coupling regime with a gap in the $\theta_1 - \theta_2$ channel. Such initial conditions correspond to $R_{1,2} > 1$, i.e., to the attractive interaction in the $\bar{\theta}_n$-channels. It is worth emphasizing here that due to the presence of inter-subband forward scattering, such effective attraction may arise in a purely repulsive system, as we shall demonstrate shortly (*case A* below). Another relevant limit is represented by the "repulsive" case (*case B*), where initially $x_0 < 0$, $y_0 < 0$. For a strong repulsion ($R_{1,2} \ll 1$), $\bar{f}$ quickly renormalizes to zero and the resulting phase is a two-subband Luttinger liquid. However, there is a region of anomalously small $x_0, y_0 \sim \bar{f}_0$ (which requires strong intersubband scattering), where Cooper scattering may still be important. One finds that if

$$(c_1 + c_2)\bar{f}_0^2 > (x_0 + y_0)^2, \tag{65}$$

the Cooper process wins over repulsion and initially negative variables x, y change sign during renormalization. In this case, the Cooper scattering amplitude initially decreases with l, but then passes through the minimum and finally flows into strong-coupling regime $\bar{f} \geq 1$. Equation (65) is a condition for the development of superconducting fluctuations in the system with purely repulsive interactions.

Now we again apply our analysis to a wire with weak and long-range interactions. Two limits are possible.

Case A: $\Delta_{CDW}/k_F \ll v_{1F} - v_{2F} \ll U_0$.
The first inequality allows one to neglect direct backscattering, which leads to inter-subband CDW, whereas the second one allows to consider subbands as "nearly equivalent". Denoting $v_F \equiv v_{1F}$, $\delta v_F \equiv v_{1F} - v_{2F}$, and the $2k_{nF}$-component of electron interaction potential in the n-th subband by $U_{2k_F}^{(n)}$, one finds

$$\frac{1}{R_1} + \frac{1}{R_2} = 2 - \frac{U_{2k_F}^{(1)} + U_{2k_F}^{(2)}}{2\pi v_F} - \frac{\delta v_F}{v_F}\frac{U_{2k_F}^{(1)}}{2\pi v_F} < 2, \tag{66}$$

which corresponds to an effective attraction. Thus, thanks to inter-subband forward scattering, the Cooper process is relevant in the system with purely repulsive interaction. Note that a small velocity imbalance $\delta v_F > 0$ enhances the relevance of Cooper scattering . (As our second subband is chosen to have a higher energy, δv_F is always positive.)

Case B: $v_{1F} - v_{2F} \gg U(0)$. Van Hove singularity.
In this limit an analytic solution is also possible. Generally, one finds that $1/R_1 + 1/R_2 > 2$, which corresponds to effective repulsion. Neither CDW nor superconducting instability can develop, and the resulting phase is a two-subband Luttinger liquid. This is not true, however, in the limit of a strong velocity imbalance, when $v_{2F}/v_{1F} \ll 1$, which corresponds to opening of the second subband for conduction. Then $u_n \approx v_n$ and $\gamma = u_1/u_2 \sim v_{1F}/v_{2F} \gg 1$. Hence $c_1 \sim \gamma$ and it follows from (65) that Cooper process wins over repulsion, if $f_C > U(0)/\sqrt{\gamma}$. For long-range interactions, this inequality reduces to

$$\frac{v_{2F}}{v_{1F}} \ll \left(\frac{a}{d}\right)^4. \tag{67}$$

The physics of this scenario is well known - interactions are enhanced due to the large value of the density of states ($\propto 1/v_{2F}$) in the upper subband (Van Hove singularity). We should warn here that our calculations do not describe the very onset of conduction in the upper subband, because its proper description requires accounting for the nonlinearity of the electron spectrum, which is beyond our bosonization analysis. However, such a calculation was performed by Balents and Fisher [15], who analyzed the case of a contact interaction. They found that superconducting fluctuations are indeed enhanced in this limit.

We thus see that superconducting fluctuations do have a good chance to overcome the electron-electron repulsion and drive the system into a strong-coupling phase with the gap in the spectrum of relative current fluctuations, $\bar{\theta}_1 - \bar{\theta}_2$.

There is another important feature of the RG-flow described by Eqs. (60-63): the interaction tries to equilibrate densities in the subbands. This is seen from equation (62): $d\gamma/dl$ is proportional to $1 - \gamma^2$, which makes $\gamma = 1$ a stable fixed point. γ tends to increase, if initially $\gamma_0 < 1$, and it tends to decrease, if $\gamma_0 > 1$.

4   Electrons with Spins

Guided by the results of the previous Section, we now comment briefly on what happens if spin is included. As should be clear from the complexity of Eqs.(16-21), this question has no simple answer. For a quantum wire with $0 < U(2k_F) < U(0)$, possible phases are (i) Luttinger liquid, (ii) inter-subband CDW, and (iii) Cooper phase (superconductor). As with spinless fermions, subbands must be nearly equivalent in order for the CDW phase to occur, whereas the Cooper phase needs effective attraction in the relative charge-density excitation channel. When neither of these conditions is met, a two-subband Luttinger liquid is realized. At different degrees of generality, renormalization group analysis of the model defined by Eqs.(16- 21) has been performed in the past and we refer to papers [25,15,17] for a detailed description.

Long-range interactions If the interaction is long-range and weak, a considerable simplification occurs. In this case, amplitudes of forward intra- and inter-subband processes are the same [see discussion after Eq.(12)] and a simple perturbative estimate of the scaling dimension δ_C of the Cooper process (21) is possible.

For weak interactions, $K_{n\rho} = 1 - (2U(0) - U_{2k_F}^{(n)})/(2\pi v_{nF})$, and $f_0 = U(0)$. The $SU(2)$-invariance requires that $K_{n\sigma} = 1$. One finds

$$\delta_C = 2 - \frac{U_{2k_F}^{(1)} + U_{2k_F}^{(2)}}{4\pi v_F} - \frac{\delta v_F}{v_F} \frac{U_{2k_F}^{(1)}}{4\pi v_F} < 2. \tag{68}$$

Observe that this result coincides with Eq.(66) upon replacing $U(0) \to 2U(0)$. Thus, Cooper scattering is relevant for repulsive long-range interactions (and assuming also that $0 < \delta v_F \ll U_0$).

58   Starykh et al.

However, if $K_{\rho-}$ is sufficiently close to its non-interacting value, i.e., to unity, backscattering is strong ($f_{bs} \gg t_{sp}, t_{tp}$), and δk_F is small, the CDW-channel can take over the Cooper one, similarly to what happened in Subsection 3.1.

Electron ladder There is an interesting qualitative question where an RG consideration is very helpful. Suppose that condition (68) is satisfied and thus tunneling of fluctuational Cooper pairs is relevant. What happens to spin excitations? To answer this question, we relax the $SU(2)$-invariance condition and perform the RG calculation for two nearly equivalent subbands so that $v_{n\nu} \equiv v_\nu, K_{n\nu} \equiv K_\nu$, where $\nu = \rho, \sigma$. Nevertheless, we assume that δk_F is still finite and neglect direct backscattering (19), similarly to Subsection 3.1. The problem then becomes identical to that of two coupled equivalent chains ("electron ladder"). Also, for the sake of simplicity, we consider only the singlet channel of Cooper scattering. Due to enhanced symmetry, the total current fluctuation mode $\theta_{\rho+} = (\theta_{1\rho} + \theta_{2\rho})/\sqrt{2}$ decouples from the rest and is described by a harmonic action with $K_{\rho+}^{-2} = K_\rho^{-2} + 2f_0/(\pi v_F)$. Relative current fluctuations $\theta_{\rho-} = (\theta_{1\rho} - \theta_{2\rho})/\sqrt{2}$ are described by the sine-Gordon theory [t_{sp} term in (21)] with $K_{\rho-}^{-2} = K_\rho^{-2} - 2f_0/(\pi v_F)$ and $v_{\rho-} = v_\rho[1 - 2f_0 K_\rho/(\pi v_F)]^{1/2}$. In addition, we have to keep track of spin-density sector, which contain the cosine term corresponding to intra-subband backscattering (17). For convenience, we denote the amplitude of this term by g_σ, its initial value being $g_\sigma(0) = U(2k_F)$. After tedious but straightforward calculations we arrive at the following system of RG equations:

$$\frac{d}{d\ell}\bar{g} = 2(1 - K_\sigma)\bar{g} - \bar{t}^2, \tag{69}$$

$$\frac{d}{d\ell}\bar{t} = (2 - K_\sigma - \frac{1}{K_{\rho-}} - \bar{g})\bar{t}, \tag{70}$$

$$\frac{d}{d\ell}(1 - K_\sigma) = \frac{1}{2}(\bar{g}^2 + \bar{t}^2), \tag{71}$$

$$\frac{d}{d\ell}(1 - \frac{1}{K_{\rho-}}) = \bar{t}^2, \tag{72}$$

where $\bar{g} = g_\sigma/(\pi v_F)$ and $\bar{t} = t_{sp}/(\pi v_F)$. Let us recall what happens in the absence of Cooper tunneling first and set $t_{sp} = 0$ everywhere in this system. In the weak-coupling limit, $K_\sigma = (1-\bar{g})^{-1/2} \approx 1+\bar{g}/2$ and (69) becomes $d\bar{g}/d\ell = -\bar{g}^2$, which gives $\bar{g}_\ell = \bar{g}_0/(1+\bar{g}_0\ell)$. For repulsive interactions ($\bar{g}_0 > 0$), $\bar{g} \propto \ell^{-1} \to 0$ as $\ell \to \infty$: intra-subband backscattering is marginally irrelevant. Observe now that when the Cooper process is present and relevant, i.e., when $\bar{t}$ increases, the flow of $\bar{g}$ is modified: the $\bar{t}^2$-term on the right-hand-side of Eq. (69) changes the sign of $d\bar{g}/d\ell$. Hence $\bar{g}_\ell$ is bound to become negative in the process of renormalization and grows unrestrictedly in its absolute value. Intra-subband spin backscattering is thus driven relevant by singlet-pair tunneling, which results in pinning of φ_σ in Eq.(17) and opening of the *spin gap*. Thus, similar to the true superconducting state in higher dimensions, the Cooper phase is characterized by gaps in both charge- and spin-channels. The only massless excitations are those of the total charge channel. This phenomenon is not restricted to the degenerate electron ladder but rather is a generic feature of the system of coupled subbands and/or chains, see, e.g., [15,17].

5 Conductance

Having realized the importance of inter-subband interactions, we now proceed with the analysis of its effect on observable properties of quantum wires. The first property we consider is conductance G.

5.1 No Disorder

Our results for the conductance of a clean wire can be understood from the following simple considerations. The dc conductance of a single-subband wire is equal to $2e^2/h$ regardless of the interactions in the wire [48–50]. Consider now a wire with several subbands occupied. Those interband interactions, which do not open gaps, lead only to a renormalization of Luttinger-liquid parameters. As these parameters do not enter the final result for G, the conductance remains quantized in units of $2e^2/h$ per occupied subband. Other processes, such as direct backscattering and Cooper scattering, open gaps in channels of *relative* charge fluctuations as well as in spin channels. Neither of these gaps, however, affects the motion of a *center-of-mass* of the electron fluid through the wire, thus G remains unrenormalized by this type of interactions as well.

Now we demonstrate the proof of the statements made above. Consider the case of a superconducting instability, when the cosine term of Cooper scattering in Eq. (12) is relevant and the $\theta_{\rho-}$-field is thus gapped. Gaussian fluctuations of the gapped field can be described by expanding the relevant cosine term around its minimum value:

$$\left(4f_C/\pi^2\alpha^2\right)\cos[\sqrt{2\pi}(\theta_{1\rho} - \theta_{2\rho})] \approx \text{const} + m^2(\theta_{1\rho} - \theta_{2\rho})^2, \qquad (73)$$

where m is the mass of the field. The strong-coupling (superconducting) phase corresponds to $m \neq 0$, whereas in the Luttinger-liquid phase $m = 0$. Expanding $\cos[\sqrt{2\pi}\varphi_{n\sigma}]$ around their minima as well, we find that at the Gaussian level charge and spin modes decouple again. Note though that now these are *massive* modes.

As spin excitations do not affect charge transport, we concentrate on the charge sector of the theory, whose Hamiltonian is given by the sum of Eqs. (16,18) and (73). In order to simplify notations, we suppress index ρ in this section, so that $\theta_{\rho-} \to \theta_-$, etc. Using $v_n K_n = v_{nF}$, we write the charge Hamiltonian as

$$H_\rho = \frac{1}{2} \int dx \Big\{ \frac{v_{1F} + v_{2F}}{2} \big[(\partial_x \theta_+)^2 + (\partial_x \theta_-)^2 \big]$$
$$+ \frac{1}{2} \Big(\frac{v_1}{K_1} + \frac{v_2}{K_2} \Big) \big[(\partial_x \varphi_+)^2 + (\partial_x \varphi_-)^2 \big]$$
$$+ \frac{2f_0}{\pi} \big[(\partial_x \varphi_+)^2 - (\partial_x \varphi_-)^2 \big] + (v_{1F} - v_{2F}) \partial_x \theta_+ \partial_x \theta_-$$
$$+ \Big(\frac{v_1}{K_1} - \frac{v_2}{K_2} \Big) \partial_x \varphi_+ \partial_x \varphi_- + m^2 \theta_-^2 \Big\}. \tag{74}$$

The total charge current is given by $j = e\sqrt{2/\pi} \sum_n \partial_t \varphi_n = e(2/\sqrt{\pi})\partial_t \varphi_+$. In order to find the current-current correlation function, one needs to know the retarded Green's function $G_{++}(xt) = -i\Theta(t)\langle[\varphi_+(xt), \varphi_+(00)]\rangle$, which is coupled to another Green's function $G_{-+}(xt) = -i\Theta(t)\langle[\varphi_-(xt), \varphi_+(00)]\rangle$ by the following equations of motion

$$(-i\partial_t)^2 G_{++} = \frac{v_1 + v_2}{2}\delta(x)\delta(t) - \frac{1}{2}\partial_x \Big\{ \frac{v_1^2}{K_1} + \frac{v_2^2}{K_2} + \frac{2f_0}{\pi}(v_1 + v_2) \Big\} \partial_x G_{++}$$
$$- \frac{1}{2}\partial_x \Big\{ \frac{v_1^2}{K_1} - \frac{v_2^2}{K_2} - \frac{2f_0}{\pi}(v_1 - v_2) \Big\} \partial_x G_{-+};$$
$$(-i\partial_t)^2 G_{-+} = \frac{v_1 - v_2}{2}\delta(x)\delta(t) - \frac{1}{2}\partial_x \Big\{ \frac{v_1^2}{K_1} + \frac{v_2^2}{K_2} - \frac{2f_0}{\pi}(v_1 + v_2) \Big\} \partial_x G_{-+}$$
$$- \frac{1}{2}\partial_x \Big\{ \frac{v_1^2}{K_1} - \frac{v_2^2}{K_2} + \frac{2f_0}{\pi}(v_1 - v_2) \Big\} \partial_x G_{++} \tag{75}$$
$$- m^2 \Big\{ \frac{1}{2}\Big(\frac{v_1}{K_1} + \frac{v_2}{K_2} \Big) - \frac{2f_0}{\pi} \Big\} G_{-+} - m^2 \frac{1}{2}\Big(\frac{v_1}{K_1} - \frac{v_2}{K_2} \Big) G_{++} .$$

In the massless limit ($m = 0$), this system of equations is solved readily. In order to model the effect of non-interacting electron reservoirs, which the wire is attached to, we assume that K_n, v_n vary with x adiabatically and approach their non-interacting values $K_n = 1, v_n = v_F$ for $x \to \pm\infty$ [48–50]. In the zero-frequency limit, the solution is particularly simple: $G_{++}(x,\omega \to 0) = 1/2i\omega$, $G_{-+}(x,\omega = 0) = 0$. As a result, the conductivity $\sigma(x,y;\omega \to 0) = (e^2/h)(2/\sqrt{\pi})^2 i\omega/2i\omega = 4e^2/h$ is x-independent, and the conductance is simply $G = 2 \times 2e^2/h$.

In order to see the effect of the gap, we consider first the case of equivalent subbands ("electron ladder"), introduced in Sec. 4. One observes immediately that conditions $v_1 = v_2$, $K_1 = K_2$ lead to complete decoupling of the equations for G_{++} and G_{-+}. As a result, the total charge mode φ_+ is not affected by the gap. Taking the boundary condition for K and v into account gives again the universal result $G = 4e^2/h$. The result for the "electron ladder" thus gives us a hint that G remains at its universal value despite the presence of the gap in the relative charge channel. In order to prove this statement in the general case, we neglect for the moment the boundary conditions for K_n and v_n, i.e., consider a uniform wire with two coupled subbands. System (75) is then solved by Fourier transformation. The key feature of the result for $G_{++}(q,\omega)$ is that it still has a pole corresponding to a massless mode $\omega \propto q$, despite the presence of the massive term. The conductivity becomes

$$\sigma(q,\omega) \sim \frac{\omega F(\omega,q)}{(\omega^2 - \bar{v}^2 q^2)(\omega^2 + \omega_m^2 - \bar{u}^2 q^2)}, \tag{76}$$

where $\bar{v}, \bar{u}$ are some renormalized velocities, ω_m is some energy proportional to m^2, and $F(\omega,q)$ is a smooth function of its arguments. As a result, $\sigma(q,\omega) = \bar{G}\delta(q)$ in the limit $\omega \to 0$, where $\bar{G}$ has a meaning of the conductance. Because we neglected the boundary conditions corresponding to presence of non-interacting leads in this calculation, $\bar{G}$ depends on all interaction parameters – v_n, K_n, and m^2–and is of course different from $4e^2/h$. It is clear though that once the boundary conditions are restored, this non-universal value is replaced by the universal factor of $4e^2/h$. The only other possibility is $G = 0$, which, however, is ruled out by the fact that $G_{++}(q,\omega)$ has a massless pole.

We thus conclude that the conductance of a clean wire remains at the universal quantized value irrespective of whether the relative charge mode is gapped or not. The case of a CDW instability can be treated in a similar manner.

5.2   Disordered Wire

A disordered two-band system in the presence of interaction-induced instabilities was considered by Orignac and Giamarchi [28] and by Egger and Gogolin [22]. Our discussion of a disordered two-subband wire follows largely these two papers.

Results of the subsequent analysis can be summarized as follows. If Cooper scattering opens a gap, the system does not become a real superconductor: at $T \to 0$ a single weak impurity splits the wire into two disconnected pieces at low enough energies and weak random potential leads to localization of electrons, similar to the case of a gapless Luttinger liquid. Nevertheless, effects of disorder are less pronounced than for a gapless Luttinger liquid. On the contrary, the CDW-state is more sensitive to disorder than a gapless Luttinger liquid.

Spinless electrons We begin by considering a single impurity described as a potential perturbation $w(x, \mathbf{r}_\perp)$. The impurity causes backscattering of electrons within the occupied subbands, as well as inter-subband backscattering. The amplitudes of the corresponding processes are given by

$$W_n(2k_{nF}) = \int dx d\mathbf{r}_\perp w(x, \mathbf{r}_\perp)\phi_n^2(\mathbf{r}_\perp) \cos(2k_{nF}x), \ n = 1, 2; \tag{77}$$

$$W_{\text{inter}} = \int dx d\mathbf{r}_\perp w(x, \mathbf{r}_\perp)\phi_1(\mathbf{r}_\perp)\phi_2(\mathbf{r}_\perp) \cos\left[(k_{1F} + k_{2F})x\right]. \tag{78}$$

If w varies slowly across the wire, then $W_{\text{inter}} \ll W_n$ due to the orthogonality of transverse wavefunctions, and we consider intra-subband backscattering first.

The bosonized form of intra-subband backscattering is

$$W_{\text{intra}}^n = \frac{W_n(2k_{nF})}{\pi\alpha}\cos[\sqrt{4\pi}\varphi_n] = \frac{W_n(2k_{nF})}{\pi\alpha}\cos[\sqrt{2\pi}(\varphi_+ \pm \varphi_-)], \ n = 1, 2;$$
(79)

so the total backscattering operator is given by

$$W_{\text{intra}} = \sum_{n=1,2} W_{\text{intra}}^n = \frac{W_2(2k_{2F}) - W_1(2k_{1F})}{\pi\alpha}\sin[\sqrt{2\pi}\varphi_+]\sin[\sqrt{2\pi}\varphi_-]$$
$$+\frac{W_2(2k_{2F}) + W_1(2k_{1F})}{\pi\alpha}\cos[\sqrt{2\pi}\varphi_+]\cos[\sqrt{2\pi}\varphi_-].$$
(80)

Note that W_{intra} is local in space and thus cannot change the RG-flows of bulk parameters of the wire. Depending on these bulk parameters, however, W_{intra} will either grow, splitting eventually the wire into two disconnected pieces, or decay, in which case the impurity effectively disappears.

(i) *Cooper phase.*
In the Cooper phase, θ_- is gapped, hence φ_- fluctuates strongly. [This follows from the fact that θ_- and φ_- are canonically conjugated fields, see Sec. 6.] On the first sight, it may seem that these strong fluctuations render W_{intra} to zero. To see that it is not so, consider a second-order impurity contribution, e.g.,

$$\left(\frac{W_1(2k_{1F})}{\pi\alpha}\right)^2 \int d\tau \int d\tau' \left(\langle e^{i\sqrt{2\pi}\varphi_-(\tau)}e^{-i\sqrt{2\pi}\varphi_-(\tau')}\rangle\right.$$
$$\times \left.\cos[\sqrt{2\pi}\varphi_+(\tau)]\cos[\sqrt{2\pi}\varphi_+(\tau')]\right]$$
(81)

As we will be explained in more details in Sec. 6, the correlator of φ_-–fields in the Cooper phase decays exponentially, i.e., $\langle e^{i\sqrt{2\pi}\varphi_-(\tau)}e^{-i\sqrt{2\pi}\varphi_-(\tau')}\rangle \sim e^{-\Delta_{SC}|\tau-\tau'|}$, where Δ_{SC} is the Cooper gap in the θ_-–channel. As a result, the double integration over τ, τ' in Eq. (81) is effectively contracted into a single one, which gives $\Delta_{SC}^{-1}(W(2k_{1F})/\pi\alpha)^2 \int d\tau \cos[\sqrt{8\pi}\varphi_+(\tau)]$. The mechanism of generating higher order impurity backscattering was discovered in [28,22]. ($\sqrt{8\pi}$ under the cosine indicates that this is a two-particle backscattering process.) Following the RG-calculations of Kane and Fisher [7], one finds that impurity backscattering, generated in this way, becomes relevant for $K_+ < 1/2$. Note that this requires rather strong electron repulsion. For weaker repulsion, impurity scales to zero and the wire retains the universal conductance of $2e^2/h$. Without superconducting correlations, i.e., when θ_- is not gapped, an impurity is effective for $K < 1$. Thus the Cooper phase weakens but does not eliminate impurity scattering.

(ii) *CDW phase.*
In this phase, $\delta k_F = 0$ and φ_- is pinned by the bulk nonlinear term $\cos[\sqrt{8\pi}\varphi_-]$, so that φ_- acquires average value $\langle\varphi_-\rangle = \sqrt{\pi/8}$. Allowing for fluctuations

around the average value, we substitute $\varphi_- = \langle \varphi_- \rangle + \delta\varphi_-$ into (80)

$$W_{\text{intra}} = \frac{W_2(2k_F) - W_1(2k_F)}{\pi\alpha} \sin[\sqrt{2\pi}\varphi_+] \cos[\sqrt{2\pi}\delta\varphi_-]$$
$$- \frac{W_2(2k_F) + W_1(2k_F)}{\pi\alpha} \cos[\sqrt{2\pi}\varphi_+] \sin[\sqrt{2\pi}\delta\varphi_-] \qquad (82)$$

Observing that for small fluctuations one can replace $\cos[\sqrt{2\pi}\delta\varphi_-] \approx 1$, we see that the first term in (82) gives the strongest contribution to backscattering, which is relevant already for $K_+ < 2$. The second term requires more work. To the second order in the amplitude of this term, an expression similar to (81) is generated, but now it involves the following average

$$\langle \sin[\sqrt{2\pi}\delta\varphi_-(\tau)] \sin[\sqrt{2\pi}\delta\varphi_-(\tau')] \rangle \sim \frac{\Delta_{CDW}}{\Delta_0} \sinh[K_0(\Delta_{CDW}|\tau - \tau'|)], \qquad (83)$$

where Δ_{CDW} is the CDW gap, Δ_0 is the ultraviolet energy cutoff, and $K_0(x)$ is the modified Bessel function, $[K_0(x) \sim e^{-x}/\sqrt{x}$ for $x \gg 1]$. This result is due to the fact that in the *massive* phase

$$\langle e^{i\sqrt{2\pi}\delta\varphi_-(\tau)} e^{\pm i\sqrt{2\pi}\delta\varphi_-(\tau')} \rangle \sim \exp\left[-K_0\left(\Delta_{CDW}/\Delta_0\right) \mp K_0(\Delta_{CDW}|\tau - \tau'|)\right]. \qquad (84)$$

As a result, correlation functions of *sines* and *cosines* of massive fields are different: the first ones decay exponentially with distance, whereas the second ones reach constant values. Thus the double integral over τ, τ' in (81) can be reduced to the single one again, and, similarly to the Cooper-phase case, two-particle impurity backscattering, relevant for $K_+ < 1/2$, is generated.

It is also straightforward to analyze the effect of inter-subband impurity scattering

$$W_{\text{inter}} = \frac{2W_{\text{inter}}(k_{1F} + k_{2F})}{\pi\alpha} \cos[\sqrt{2\pi}\varphi_+] \cos[\sqrt{2\pi}\theta_-]. \qquad (85)$$

Similarly to Eq.(81), we have to average over the strongly fluctuating θ_-–field, which generates again the two-particle backscattering term $\sim \cos[\sqrt{8\pi}\varphi_+]$, relevant for $K_+ < 1/2$.

Hence, the perturbative correction to the conductance of a CDW wire behaves as

$$-\delta G_{CDW} \propto w^2 \epsilon^{K_+ - 2} + \left(w^2/\Delta_{CDW}\right)^2 \epsilon^{4K_+ - 2}, \qquad (86)$$

where $\epsilon = \max\{T, \text{bias}\}$ and where we have also indicated the order of the impurity potential. Please note that the exponent of the weak-link counterpart of (86), derived in Eq.(134) of Section 7, is not related to the leading exponent $K_+ - 2$ here by the conventional duality relation [7]. We conjecture that this violation of the duality signals phase transition separating regimes of weak and strong tunneling.

For the Cooper-phase case, $\cos[\sqrt{2\pi}\theta_-]$ is replaced by $\sin[\sqrt{2\pi}\delta\theta_-]$. Hence, W_{inter} also generates the effective two-particle term in the second order of perturbation theory (cf. with our analysis of the second term in (82)). The correction to the conductance is given by

$$-\delta G_{SC} \propto \left(w^2/\Delta_{SC}\right)^2 \epsilon^{4K_+-2}. \tag{87}$$

To summarize, the Cooper phase is insensitive to a single impurity as long as $K_+ > 1/2$, whereas the CDW one is stable only for $K_+ > 2$.

We now turn to the case of weak random potential produced by many impurities. To establish the boundary between delocalized and localized regimes for the case of weak disorder, it suffices to replace $\epsilon \to 1/L$ in Eqs. (86,87) and to multiply δG by the total number of impurities [51], proportional to L. Depending on whether δG increases or decreases with L, the wire is in the localized or delocalized phase. By doing so, one concludes that a wire is localized for $K_+ < 3$, if it is in the CDW phase, and for $K_+ < 3/4$, if it is in the Cooper phase. For comparison, a (spinless) single-subband Luttinger liquid is localized for $K < 3/2$.

Electrons with spins The bosonized form of impurity backscattering is given by

$$W_{\text{intra}}^{1(2)} = \frac{4W_{1(2)}(2k_{F1,2})}{2\pi\alpha} \cos[\sqrt{\pi}(\varphi_{\rho+} \pm \varphi_{\rho-})] \cos[\sqrt{\pi}(\varphi_{\sigma+} \pm \varphi_{\sigma-})]; \tag{88}$$

$$W_{\text{inter}} = \frac{4W_{\text{inter}}(k_{1F} + k_{2F})}{2\pi\alpha} \cos[\sqrt{\pi}(\varphi_{\rho+} + \theta_{\rho-})] \cos[\sqrt{\pi}(\theta_{\sigma-} + \varphi_{\sigma+})], \tag{89}$$

where $+(-)$ in the argument of cosines refers to the 1st (2nd) subband and all operators are evaluated at the position of the impurity.

In the Cooper phase, the $\theta_{\rho-}$ and $\varphi_{\sigma\pm}$–modes are gapped, whereas the conjugated modes, i.e., $\varphi_{\rho-}$ and $\theta_{\sigma\pm}$, exhibit strong fluctuations. Integrating out $\varphi_{\rho-}$ and $\varphi_{\sigma\pm}$, we find: $W_{\text{intra}} \sim \Delta_{SC}^{-1} \left[W(2k_{1,2\ F})\right]^2 \cos[\sqrt{4\pi}\varphi_{\rho+}]$, which is relevant for $K_{\rho+} < 1$. The same is true for W_{inter}, where strong fluctuations of $\theta_{\sigma-}$ produce a similar operator. The correction to the conductance behaves as

$$-\delta G_{SC} \propto \left(w/\Delta_{SC}\right)^2 \epsilon^{2K_{\rho+}-2}, \tag{90}$$

i.e., as if we were dealing with a single channel gapless Luttinger liquid, characterized by parameter $K_{\rho+}$, subject to an effectively reduced impurity potential. Weak random potential leads to localization for $K_{\rho+} < 3/2$.

In the CDW-state, the situation is different. In this case, the $\varphi_{\rho-}$- and $\varphi_{\sigma\pm}$-modes are gapped [14,28], whereas the $\theta_{\rho-}$- and $\theta_{\sigma\pm}$- modes fluctuate strongly. As a result, intersubband backscattering is renormalized into $\cos[\sqrt{4\pi}\varphi_{\rho+}]$, as in the Cooper phase, but intra-subband one remains unchanged and is determined by the dynamics of the only gapless $\varphi_{\rho+}$ mode:

$$W_{\text{intra}} \propto \cos[\sqrt{\pi}\varphi_{\rho+}]. \tag{91}$$

Pinning of $\varphi_{\rho+}$ at the impurity site leads to the suppression of the conductance. W_{intra} is relevant, i.e., an impurity eventually splits the wire into two disconnected wires for $K_{\rho+} < 4$. The correction to the conductance behaves as $-\delta G \propto w^2 \epsilon^{\frac{K_{\rho+}}{2}-2}$. This is to be contrasted with the case of a gapless Luttinger liquid, when the impurity is relevant only for repulsive interactions ($K < 1$). This reflects the fact that a real (gapped) charge-density-wave is pinned stronger than the fluctuating one (Luttinger liquid). Finally, a weak random potential localizes the CDW-wire for $K_{\rho+} < 6$. It is worth pointing out here that such large values of critical $K_{\rho+}$, separating localized from pure behavior, imply strong effective attraction between charge fluctuations in the $\rho+$ channel. It might well be that for such an attraction instead of CDW instability charge segregation will take place [52].

Thus, for electrons with spin, the Cooper phase is more stable to impurities than the CDW one, similar to a spinless case, but both are unstable in the physically relevant region of $K_{\rho+} < 1$.

6   Single-Particle Density of States

We now turn to the discussion of tunneling into a quantum wire. To the leading order in barrier transparency $\mathcal{T}$, the differential tunneling conductance is

$$G(V) = \frac{dI}{dV} = |\mathcal{T}|^2 \rho_c \rho(eV), \qquad (92)$$

where ρ_c is the density of states (DOS) in the contact (which we assume to be energy independent), $\rho(\epsilon)$ is the DOS of the wire, and V is the applied voltage. When the wire is in the gapless Luttinger-liquid phase, $\rho(\epsilon) \propto |\epsilon|^\beta$. This behavior has recently been observed in tunneling into carbon nanotubes [3]. Tunneling into the edge of a fractional quantum Hall system also exhibits a power-law current-voltage dependence [53], which might be an indication of a chiral Luttinger-liquid state at the edge. Here, however, the situation is not that straightforward, and other explanations, different from a chiral Luttinger liquid, have also been suggested [54,55].

Suppose now that a two-subband quantum wire is in one of the possible gapped phases, i.e., CDW or Cooper phase. The goal of this Section is to analyze what would a tunneling experiment show in this case. The answer turns out to depend crucially on the geometry of the experiment. If the tunneling contact probes the interior of the wire, the gapped behavior is predicted: $G(V) = 0$ for $eV < \Delta$, where Δ is the appropriate energy gap. For $eV > \Delta$ the behavior is non-universal: the threshold behavior of ρ is determined by gapless charge and spin modes. More surprisingly, tunneling into the end of the CDW-wire exhibit a gapless behavior, similar to the Luttinger-liquid case. The tunneling exponent though is different from that for the gapless phase.

6.1   Tunneling Preliminaries

The local single-particle (or tunneling) density of states is given by

$$\rho(\omega, x) = -\frac{1}{\pi} Im\{G_{ret}(\omega, x)\}, \qquad (93)$$

where $G_{ret}(\omega, x)$ is a Fourier transform of retarded Green's function $G_{ret}(t, x) = -i\Theta(t)\Sigma_s\langle\{\Psi_s(t, x), \Psi_s^\dagger(0, x)\}\rangle$. Representing the electron of the n-th subband as a sum of right- and left-movers and accounting for the orthogonality of transverse wavefunctions, the Green's function becomes $G_{ret}(t, x) = \sum_n[G_{ret}^{R_n}(t, x) + G_{ret}^{L_n}(t, x)]$, where the summation is over all occupied subbands and $G_{ret}^{N_n}(t, x) = -i\Theta(t)\ \Sigma_s\langle\{N_{ns}(t, x), N_{ns}^\dagger(0, x)\}\rangle$, $N = L, R$, see (13,14). The contribution of the off-diagonal terms $\sim \langle R_n L_n^\dagger\rangle$ is less singular and is thus neglected. $\rho(\omega, x)$ is a sum of contributions from right and left movers of all occupied subbands. To find $\rho(\omega, x)$, it is convenient to calculate first the Matsubara Green's function

$$G^R(\tau, x) = -\langle T_\tau R_s(\tau, x) R_s^\dagger(0, 0)\rangle, \qquad (94)$$

and then to make the analytic continuation to real frequencies. Left- and right-moving fermions give identical contributions to ρ, thus the result obtained from (94) is simply multiplied by a factor of two at the end.

The key feature of gapped phases in multisubband 1D systems is the co-existence of gapped and gapless modes, which also makes the calculations to be slightly less trivial. The single-particle Green's function under these circumstances has recently been considered in Ref. [27,26], and our analysis follows largerly these two papers.

6.2   Warm-Up: DOS of a Half-Filled Hubbard Chain

To warm up, we consider the simplest system in which gapped and gapless modes co-exist – a single-band Hubbard chain at half-filling. In the context of nanostructure physics, such a system is produced by imposing an artificial periodic potential of period a_0 over a quantum wire [31]. At half-filling, the Fermi-momentum $k_F = \pi/2a_0$ is commensurate with the reciprocal lattice spacing, which gives rise to Umklapp scattering. An Umklapp process occurs as simultaneous backscattering of two right- or left-moving electrons, the total momentum transferred to the lattice being $\pm 4 \times \pi/2a_0 = \pm 2\pi/a_0$. This process is responsible for opening of the (Mott-Hubbard) gap in the charge excitations spectrum. On the other hand, spin excitations remain gapless, and are described by the $SU(2)$-invariant Luttinger-liquid Hamiltonian ($K_\sigma = 1$).

The corresponding Hamiltonian of the charge sector is

$$H = \frac{1}{2}\int dx \{v_\rho K_\rho(\partial_x\theta_\rho)^2 + \frac{v_\rho}{K_\rho}(\partial_x\varphi_\rho)^2 + g\cos[\sqrt{8\pi}\varphi_\rho]\}, \qquad (95)$$

where the last (cosine) term represents Umklapp scattering. Substituting Eq. (13) into Eq. (94), one finds that G^R factorizes into a product of spin and charge parts

$$G^R(\tau, x) = -\frac{sgn(\tau)}{2\pi\alpha} F_\sigma(\tau, x) F_\rho(\tau, x);$$

$$F_\nu = \langle \exp[i\sqrt{\frac{\pi}{2}}(\varphi_\nu(1) - \varphi_\nu(0))] \exp[-i\sqrt{\frac{\pi}{2}}(\theta_\nu(1) - \theta_\nu(0))]\rangle_\nu; \nu = \rho, \sigma \quad (96)$$

where shorthand notations $1 \equiv (\tau, x)$, $0 \equiv (0, 0)$ have been used. F_σ is gapless, whereas F_ρ contains massive fields.

Bosonic calculation To calculate F_ρ in the bosonic language, we adopt the semiclassical approximation, i.e., expand the cosine in (95) around its minimum to the second order in fluctuations. This is equivalent to replacing $g\cos[\sqrt{8\pi}\varphi_\rho] \to m^2\varphi_\rho^2$, which defines the mass $m^2 \equiv 4\pi g$. Now the averaging is straightforward:

$$F_\rho(\tau, x) = \quad (97)$$

$$\exp\left\{-\frac{\pi}{2}\int \frac{d^2k}{(2\pi)^2}(1 - \cos[\boldsymbol{k}\cdot\boldsymbol{z}])\left(G_\theta(k) + G_\varphi(k) + 2iG_\theta(k)\frac{k_0k_1}{vKk_1^2 + m^2}\right)\right\},$$

where $\boldsymbol{k} \equiv (k_0, k_1) = (\omega_n, q)$, $\boldsymbol{z} \equiv (\tau, x)$ and

$$G_\varphi(\omega_n, q) \equiv \langle T_\tau \varphi_\rho\varphi_\rho \rangle_{\omega_n q} = \frac{vK}{v^2q^2 + \omega_n^2 + m^2vK}. \quad (98)$$

The Green's function of θ_ρ–fields can be written as

$$G_\theta(\omega_n, q) \equiv \langle T_\tau \theta_\rho\theta_\rho \rangle_{\omega_n q} = \frac{v}{K}\left(\frac{1}{v^2q^2 + \omega_n^2} + \frac{\omega_n^2}{v^2q^2}\frac{\bar{m}^2}{(v^2q^2 + \omega_n^2)(v^2q^2 + \omega_n^2 + \bar{m}^2)}\right)$$

$$= G_\theta^{(1)}(\omega_n, q) + G_\theta^{(2)}(\omega_n, q), \quad (99)$$

where $\bar{m}^2 \equiv m^2vK$. The first term in Eq. (99) is just a free Green's function, whereas the second one is present only in strong-coupling phase and contains a strong infrared divergence at $q \to 0$. This divergence is often explained by the "uncertainty principle": in a gapped phase, a position-like field (φ_ρ) acquires an average value, hence its canonical conjugate, momentum-like field (θ_ρ) fluctuates strongly, hence its correlation function diverges. Let us analyze the Fourier transform of G_θ in more details, and define

$$I(x, \tau) = \int \frac{d^2k}{(2\pi)^2}(1 - \cos[\boldsymbol{k}\cdot\boldsymbol{z}])G_\theta^{(2)}(\omega, q). \quad (100)$$

Changing to polar coordinates $\omega_n = k\cos\phi, q = k\sin\phi, \tau = z\cos\vartheta, x = z\sin\vartheta$, we get

$$I(z, \vartheta) = \frac{\bar{m}^2z^2}{(2\pi)^2}\int_0^{2\pi} d\phi\frac{\cos^2\phi}{\sin^2\phi}\cos^2(\phi - \vartheta)\int_0^\infty \frac{dy}{y}\frac{1 - \cos y}{y^2 + \bar{m}^2z^2\cos^2(\phi - \vartheta)}. \quad (101)$$

The integral over ϕ diverges at $\phi = 0$, which, if taken literally, means that $I = \infty$ and thus $F_\rho(\tau, x = 0) = 0$ for any finite τ. However, this divergence is absent at $\vartheta = \pi/2$, which corresponds to $\tau = 0, z = x$. Let us therefore continue the calculation at this special point. Despite the cancellation of the infrared divergence, the integral is still controlled by the region of small ϕ: $\phi \sim 1/(\bar{m}x) \ll 1$. Expanding $\sin \phi \sim \phi$ and extending the limits of angular integration to $\pm\infty$, we find that $I(z, \vartheta = \pi/2) = \bar{m}x/4 + O(1/(\bar{m}x))$ for $\bar{m}x \gg 1$. Collecting regular contributions from G_φ and $G_\theta^{(1)}$, we find that the equal-time exponential correlator of θ_ρ–fields is given by

$$\langle e^{ia\theta_\rho(x,0)} e^{-ia\theta_\rho(0,0)} \rangle = \exp\left[-\frac{a^2}{4\pi K} \ln(x^2/\alpha^2) - \frac{a^2}{4K}\bar{m}x \right]. \qquad (102)$$

An important feature here is that the expected exponential decay of this correlator is modified by the power-law prefactor, given by the usual Luttinger-liquid correlator. Symbolically, $\langle e^{ia\theta_\rho(x,0)} e^{-ia\theta_\rho(0,0)} \rangle_{\bar{m}\neq 0} = \langle e^{ia\theta_\rho(x,0)} e^{-ia\theta_\rho(0,0)} \rangle_{\bar{m}=0}$ $\times e^{-\bar{m}x}$. One should be careful in using Eq.(102): the Luttinger-liquid parameter K, which appears here, should in fact be understood as the strong-coupling fixed-point value, K^*, which is often unknown.

Fortunately, the fixed-point value of K is known for a half-filled Hubbard chain: $K^* = 1/2$. For $\bar{m}x \gg 1$ then

$$\langle e^{i\sqrt{\frac{\pi}{2}}\varphi_\rho(x)} e^{-i\sqrt{\frac{\pi}{2}}\varphi_\rho(0)} \rangle \to const,$$
$$\langle e^{i\sqrt{\frac{\pi}{2}}\theta_\rho(x)} e^{-i\sqrt{\frac{\pi}{2}}\theta_\rho(0)} \rangle \sim \exp[-\pi\bar{m}x/4]/\sqrt{x}, \qquad (103)$$

and the full Green's function behaves as

$$G^R(0, x) \propto \frac{\exp[-\pi\bar{m}x/4]}{\sqrt{x}} \times \frac{1}{\sqrt{x}}, \qquad (104)$$

in agreement with Ref. [57]. The second $x^{-1/2}$-factor in Eq. (104) is due to gapless spin excitations.

So far all calculations have been straightforward. Now we would like to argue that the infrared divergence of $I(z, \vartheta \neq 0)$ is an artifact of the semiclassical approximation, which ignores degeneracy of $\cos[\sqrt{8\pi}\varphi]$ with respect to a uniform shift $\varphi \to \varphi + \sqrt{\frac{\pi}{2}}N$ with integer N. The proper theory of both massive and massless phases should be Lorentz-invariant. We thus propose that the correct result, valid for any $z = \sqrt{x^2 + v^2\tau^2}$, is given by Eq.(102) where x is replaced by Euclidian distance z: $x \to z$. Therefore,

$$\langle e^{ia\theta(x,\tau)} e^{-ia\theta(0,0)} \rangle = \exp\left[-\frac{a^2}{4\pi K} \ln\left(\frac{x^2 + v^2\tau^2}{\alpha^2}\right) - \frac{a^2}{4K}\bar{m}\sqrt{x^2 + v^2\tau^2} \right]. \qquad (105)$$

Similar arguments in favor of such replacement were given by Voit [26].

We now use Eq. (105) to evaluate Eq. (97), and find [compare with (103)]

$$F_\rho(\tau, 0) \sim \sqrt{\frac{\alpha}{v|\tau|}} \exp[-\pi\bar{m}v|\tau|/4]. \qquad (106)$$

The spin sector average is non-zero and universal (thanks to $K_\sigma = 1$), therefore $F_\sigma(\tau, 0) \sim \sqrt{\alpha/v|\tau|}$. The corresponding DOS will be calculated later, see Eq.(117). The correctness of the procedure described above is verified in the next Section.

Re-fermionization. To check that our proposition makes sense, we now switch gears and derive Eq. (106) in a completely different way. To this end, we use the Luther-Emery refermionization procedure [58], which works for $K = 1/2$, that is at the fixed-point of a half-filled Hubbard chain. This procedure begins with an innocuous looking transformation $\varphi_\rho = \varphi/\sqrt{2}$, $\theta_\rho = \sqrt{2}\theta$, which changes Umklapp scattering in Eq.(95) into backscattering of some auxiliary particles (solitons): $\cos(\sqrt{8\pi}\varphi_\rho) = \cos(\sqrt{4\pi}\varphi)$. Right- and left-going solitons are defined by

$$\psi_\pm = \frac{1}{\sqrt{2\pi\alpha}} \exp\left\{\pm i\sqrt{\pi}(\varphi \mp \theta)\right\}. \tag{107}$$

In terms of new bosons, the original fermion operator (13) becomes

$$R_s = \frac{e^{-i\pi/8}}{\sqrt{2\pi\alpha}} \exp[is\sqrt{\frac{\pi}{2}}(\varphi_\sigma - \theta_\sigma)] \, \exp(i\frac{\sqrt{\pi}}{2}\varphi) \exp(-i\sqrt{\pi}\theta). \tag{108}$$

It can also be written in terms of solitons

$$R_s = e^{i\pi/8} \exp[is\sqrt{\frac{\pi}{2}}(\varphi_\sigma - \theta_\sigma)] \exp(-i\frac{\sqrt{\pi}}{2}\varphi) \, \psi_+, \tag{109}$$

where φ is expressed in terms of soliton density fluctuations as $(1/\sqrt{\pi})\partial_x\varphi =:$ $\psi_+^\dagger\psi_+ + \psi_-^\dagger\psi_-$:. The usefulness of these formal manipulations is based on the fact that Hamiltonian (95) is quadratic in *massive* solitons $\psi_\pm$, and the mass (or the gap Δ) in their spectrum is determined by coupling constant g: $\Delta = g/(2\pi\alpha)$. Due to the presense of the gap, charge density fluctuations are suppressed, which results in the suppression of fluctuations of φ. Therefore, at energies below the gap the phase factor $\exp(-i\sqrt{\pi}\varphi/2)$ in Eq. (109) can be replaced by its average value. The charge part of the Green's function (F_ρ) coincides with the Green's function of massive fermions

$$F_\rho(\tau, x) \sim \int \frac{d\omega_n dq}{(2\pi)^2} e^{i\omega_n\tau + iqx} \left(-\frac{i\omega_n + vq}{\omega_n^2 + \Delta^2 + v^2q^2}\right). \tag{110}$$

At $x = 0$,

$$F_\rho(\tau, 0) \sim -\frac{1}{v}\int d\omega_n \frac{i\omega_n e^{i\omega\tau}}{\sqrt{\omega_n^2 + \Delta^2}} = -\frac{1}{v}\partial_\tau \int d\omega_n \frac{\cos(\omega_n\tau)}{\sqrt{\omega_n^2 + \Delta^2}}$$

$$= \frac{\Delta}{v} sgn(\tau) K_1(\Delta |\tau|). \tag{111}$$

Asymptotically, $F_\rho(\tau, 0) \sim (\Delta/v\sqrt{\Delta|\tau|})e^{-\Delta|\tau|}$, in agreement with our earlier proposition (106).

It follows from Eqs.(110)-(111) that upon continuing to real frequencies $F_\rho^{ret}(\omega = i\omega_n) = -(\omega/2v)(\Delta^2 - \omega^2)^{-1/2}$. Hence, the DOS of massive fermions is given by

$$\rho_\rho(\omega, x = 0) = \frac{1}{2\pi v}\Theta(\omega - \Delta)\frac{\omega}{\sqrt{\omega^2 - \Delta^2}}. \tag{112}$$

Up to a factor of $1/2$, which is due to the fact that this is the contribution of right-movers only, the obtained result is just the DOS of free massive particles with dispersion $\epsilon(q) = \sqrt{v^2 q^2 + \Delta^2}$. Since there are no particles above energy Δ at zero temperature, there are no interaction corrections to the density of states as well [59]. Another way of deriving this result consists in using the Ising-model representation of φ, θ fields (see Ref. [60] for details). In this representation,

$$\exp(-i\sqrt{\pi}\theta) = \sigma_1\mu_2 - i\mu_1\sigma_2,$$
$$\exp(i\sqrt{\pi}\varphi) = \mu_1\mu_2 + i\sigma_1\sigma_2, \tag{113}$$

where μ_i (σ_i) are order (disorder) fields of the $d = 2$ Ising model, whose correlators are known. At long times, i.e., when $\Delta\tau \gg 1$,

$$\langle\mu_i(\tau)\mu_j(0)\rangle \sim \delta_{ij}\langle\mu\rangle^2,$$
$$\langle\sigma_i(\tau)\sigma_j(0)\rangle \sim \delta_{ij}K_0(\Delta|\tau|). \tag{114}$$

As a result, $F_\rho(\tau) \sim sgn(\tau)K_0(\Delta|\tau|)$. Because of the condition $\Delta\tau \gg 1$, there is no discrepancy between Eqs. (114) and (111), since the leading asymptotic term of $K_\nu(x)$ is ν-independent. Hence one again finds a square-root singularity in $F_\rho^{ret}(\omega)$ for $\omega - \Delta \ll \Delta$. The correspondence with the Ising model allows one to estimate neglected terms as $e^{-3\Delta\tau}$.

A more illuminating way to understand the square-root singularity is provided by the real-time calculation. Starting from Eq.(111), it can be shown that [61]

$$Im[F_\rho^{ret}(\omega)] = \frac{\Delta}{4\pi v}\int_{-\infty}^{\infty} dt e^{i\omega t}\left(K_1(-i\Delta t) - K_1(i\Delta t)\right)$$
$$= \frac{\Delta}{2v}\int_0^{\infty} dt \sin(\omega t)Y_1(\Delta t) = \Theta(\omega - \Delta)\frac{\omega}{2v\sqrt{\omega^2 - \Delta^2}}, \tag{115}$$

where $Y_1(x)$ is the Bessel function of the second kind. Since asymptotically $Y_1(x) \sim \sin(x)/\sqrt{x}$, the origin of the singularity at $\omega = \Delta$ can be easily understood. For $\omega - \Delta \ll \Delta$, the integrand of (115) oscillates very slowly, with period $t_0 = 2\pi/(\omega - \Delta)$. The integral is thus determined by long times, $t \approx t_0$, and can be estimated as $\int_0^{t_0} 1/\sqrt{t} \sim \sqrt{t_0}$. We see that the threshold behavior of the DOS is determined by times much longer than $1/\Delta$, which justifies our use of the long-time asymptotics of Bessel functions to evaluate the DOS at $\omega \approx \Delta$.

DOS of a physical electron To find the density of states of a physical electron, we have to convolute Eq.(112) with the contribution of the gapless spin mode:

$$\rho(\omega, x = 0) = \int \frac{d\epsilon_n}{2\pi} F_\rho(\epsilon_n, x = 0) F_\sigma(\omega_n - \epsilon_n, x = 0)|_{\omega_n = -i\omega}$$

$$= \frac{2}{\pi} \int_0^\omega d\epsilon Im[F_\rho^{ret}(\epsilon)] Im[F_\sigma^{ret}(\omega - \epsilon)]. \tag{116}$$

Since $F_\sigma(\epsilon_n) \sim \sqrt{\alpha/iv\epsilon_n}$, we find

$$\rho(\omega, x = 0) = \frac{2}{\pi v} \sqrt{\frac{\alpha}{v}} \Theta(\omega - \Delta) \int_\Delta^\omega d\epsilon \frac{\epsilon}{\sqrt{\epsilon^2 - \Delta^2}} \frac{1}{\sqrt{\omega - \epsilon}}. \tag{117}$$

For $\omega - \Delta \ll \Delta$, where our derivation is valid,

$$\rho(\omega, x = 0) = \frac{\pi}{v} \sqrt{\frac{g}{v}} \, \Theta(\omega - \Delta). \tag{118}$$

Instead of a square-root singularity Eq. (112), the DOS of a physical electron exhibits a regular behavior approaching a finite value at the threshold. This modification is due to dressing of the gapped charge mode by gapless spin excitations. At energies much above the gap DOS increases, $\rho(\omega, x = 0) \sim \sqrt{\omega}$, which means that the spectral weight is shifted to higher energies. The energy-independent electron DOS near the threshold was obtained in [26,27]. Parenthetically, the functional form (118) remains valid when the spin channel is gapped as well. In this case the density of states of spin excitations is given by Eq.(112) with $\Delta \to \Delta_\sigma$. Doing the integral (116), we again find behavior (118) near the threshold $\Delta + \Delta_\sigma$. In other words, the gap in the electron's DOS is given by the sum of charge- and spin-sector gaps.

6.3 DOS of the Electron Ladder

We consider now the tunneling density of states of a two-channel wire. Application of Eq.(102) requires knowledge of strong-coupling fixed point values of K_ν, which are not known for a general case of two non-equivalent channels coupled by the Coulomb interaction. Just to illustrate what kind of behavior one might expect in this case, we consider an electron ladder with $K_{1\nu} = K_{2\nu}$ ($\nu = \rho, \sigma$) in the Cooper phase. The right-moving fermion is represented by

$$R_{n=1,s} = \frac{e^{ik_F x}}{\sqrt{2\pi\alpha}} e^{is\sqrt{\pi}(\varphi_{\sigma-} - \theta_{\sigma-})/2} e^{is\sqrt{\pi}(\varphi_{\sigma+} - \theta_{\sigma+})/2}$$

$$\times e^{i\sqrt{\pi}(\varphi_{\rho-} - \theta_{\rho-})/2} e^{i\sqrt{\pi}(\varphi_{\rho+} - \theta_{\rho+})/2}. \tag{119}$$

Now we apply Eq.(102) to the correlator $F(\tau) = -\langle T_\tau R_{1,s}(\tau, 0) R_{1,s}^\dagger(0, 0)\rangle$. In the Cooper phase, $\theta_{\rho-}$ and $\varphi_{\sigma\pm}$ are gapped, hence their conjugates are exponentially suppressed. As a consequence, e.g.,

$$\langle e^{i\sqrt{\pi}(\varphi_{\sigma-}(\tau) - \theta_{\sigma-}(\tau))/2} e^{-i\sqrt{\pi}(\varphi_{\sigma-}(0) - \theta_{\sigma-}(0))/2}\rangle = C_\sigma \left(\frac{\alpha}{v_\sigma |\tau|}\right)^{1/8} \exp\left[-\frac{\pi m_\sigma v_\sigma |\tau|}{16}\right]. \tag{120}$$

On the other hand $\varphi_{\rho+}$ and $\theta_{\rho+}$ remain critical. As a result,

$$F(\tau) \propto sgn(\tau) \, |\tau|^{-\kappa} \exp\left[-\frac{\pi}{16}(2m_\sigma v_\sigma + m_{\rho-} v_{\rho-})|\tau|\right], \quad \kappa = \frac{1}{8}(3 + K_{\rho+} + 1/K_{\rho+}).$$
(121)

Hence the DOS behaves as

$$\rho(\omega) \propto \frac{\Theta(\omega - \Delta_{SC})}{(\omega - \Delta_{SC})^\gamma}, \quad \gamma = 1 - \kappa = \frac{1}{8}(5 - K_{\rho+} - 1/K_{\rho+}), \qquad (122)$$

where $\Delta_{SC} = (\pi/16)(2m_\sigma v_\sigma + m_{\rho-} v_{\rho-})$. Note that $\gamma \le 3/8 < 1/2$ for $K_{\rho+} \le 1$.

Exactly at half-filling, when the $\rho+$ mode is also gapped due to Umklapp scattering and there are no more gapless modes, our procedure gives $F(\tau) \sim e^{-\Delta|\tau|}/\sqrt{|\tau|}$, in agreement with recent exact result [62]. The corresponding DOS is that of a free massive particle, $\rho_{hf}(\omega) \sim \Theta(\omega - \Delta)/\sqrt{\omega - \Delta}$.

Comparison of (122) with (118) shows that softening of the square-root singularity is less pronounced for the electron ladder than for the Hubbard chain, because there are three gapped and only one gapless mode now. As repulsion in the $\rho+$ channel becomes stronger, i.e., as $K_{\rho+}$ decreases, the singularity becomes weaker and disappears at $K_{\rho+} = \left(5 - \sqrt{21}\right)/2 \approx 0.2$. For even smaller $K_{\rho+}$, we have $\rho(\omega = \Delta_{SC}) = 0$. This behavior, though, is not very realistic as it requires very strong repulsion. For weak repulsion, i.e., when $K_{\rho+} \approx 1$,

$$\rho(\omega) \propto \Theta(\omega - \Delta_{SC})/(\omega - \Delta_{SC})^{3/8}, \qquad (123)$$

and the threshold singularity is still present albeit softened compared to the free massive particle case. We note that $\rho(\omega)$ (122) is similar to the DOS of high unoccupied subbands of the wire, considered recently by Balents [63].

One would expect that the long-range order of the Cooper phase affect tunneling. Indeed one finds that *pair correlations* are determined by the Luttinger parameter of the total charge fluctuations only [18],

$$\langle R_{1,s}(\tau) L_{1,-s}(\tau) \, L^\dagger_{1,-s}(0) R^\dagger_{1,s}(0) \rangle \sim \langle e^{-i\sqrt{\pi}\theta_{\rho+}(\tau)} e^{i\sqrt{\pi}\theta_{\rho+}(0)} \rangle \sim \tau^{-1/(2K_{\rho+})},$$
(124)

whereas all other two-particle combinations decay exponentially. Thus, although the single-particle density of states is strictly zero, the two-particle one is not. In principle, this effect can be checked experimentally by tunneling into a two-channel wire from the superconducting tip - one should observe then a nonzero tunneling current of Cooper pairs. Its magnitude, however, will be much smaller than the current in a system of a normal tip and gapless wire, because the probability of two-particle (Copper pair) tunneling ($|\mathcal{T}|^4$) is exponentially smaller than that of single-particle tunneling ($|\mathcal{T}|^2$).

7   Tunneling into the End of a Gapped Wire

Tunneling into the end of a Luttinger liquid is different from tunneling into the bulk. The reason for this difference is the open boundary condition $\psi = 0$

for the electron wavefunction. For the boson modes describing charge and spin displacements, this condition means pinning at the boundary. The difference between the edge and bulk tunneling was considered first theoretically by Kane and Fisher [7], and has been recently observed in experiments with the tuneling into a carbon nanotube [3]. A rigorous treatment of a Luttinger liquid with open boundary conditions, which involves re-formulation of the bosonization procedure, can be found in Refs. [64–66].

Suppose now that a two-subband wire is driven into a CDW state by direct backscattering processes accompanied by density adjustment, as described in Sec. 3.1. The relative mode of charge excitations is described by Hamiltonian (46), in which we put $\delta k_F = 0$. From the equivalence of Eqs. (95) and (46) (with $\delta k_F = 0$), we expect the fixed point-value of K_- in the CDW-phase to be the same as for a half-filled Hubbard chain, i.e., $K_-^* = 1/2$. The total charge mode (φ_+) remains gapless and plays the same role as the spin mode of the Hubbard chain [see Eqs. (118) and (122)]: it softens the threshold singularity of the DOS. For $\omega < \Delta_{CDW}$, the density of states is equal to zero. Thus if the tunneling contact probes the interior of the wire, a gapped behavior is observed.

However, the DOS at the end of a wire exhibits *gapless* behavior, i.e, $\rho_{end} \propto |\omega|^{\alpha_d}$, as we will demonstrate in the rest of this Section. Consequently, $I(V) \propto |V|^{\alpha_d+1}$ for tunneling from a macroscopic (Fermi-liquid contact) into the end of the wire, and $I(V) \propto |V|^{2\alpha_d+1}$ for tunneling through a barrier located somewhere within the wire.

This very different behavior of the DOS in the bulk and at the end of the wire can be understood physically for tunneling through a barrier located within the wire (cf. Fig. 4). Without the barrier, the CDW is free to slide and the conductance is the same as in the absence of any interactions. Squeezing one electron into the middle of the wire leads to creation of a soliton-like compression in one of the two modes, and to accompanying it "stretch" in the other mode, which requires an energy of the order of the charge gap Δ_{CDW}. However, such an excitation needs not be created when the barrier distorts the uniform profile of the CDW. Indeed, the boundary condition imposed by the barrier pins the mode φ_- at $x = 0$ to a value which is different from the one in the bulk, $\varphi_-(x \to \infty) = \sqrt{\pi/8}$ (the latter follows from the minimization of the CDW energy, as illustrated in Fig.5). Therefore, the regular order of CDW is already frustrated near the barrier: there is a built-in compression in one of the modes, and depression in the other. The electron that tunnels through the barrier arrives into the "stretched"mode. Upon the proper shift of both modes, the system arrives into a state with the same energy but with a switched "polarity"of frustration. This consideration is true if the barrier is strong enough to destroy the CDW order in its vicinity, i.e., the barrier height is larger than Δ_{CDW}.

We now give a derivation for the current through the barrier. To model the boundary conditions corresponding to a wire cut into two semi-infinite pieces, we choose the potential barrier in the form $w(x, \mathbf{r}_\perp) = \pi\alpha W \delta(x)$, so that $W_{n=1,2} = \pi\alpha W$ and $W_{inter} = 0$ [cf. Eqs. (77,78)]. To find the current in, e.g., the 1st

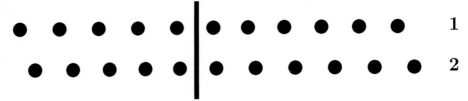

Fig. 5. Illustration of the CDW in the vicinity of a barrier : while the bulk value for φ_- is $\sqrt{\pi/8}$, the barrier forces φ_- to vanish at $x = 0$.

subband, through the barrier at $x = 0$

$$I \propto \lim_{t \to \infty} \langle \partial_t \varphi_1(x = 0, t) \rangle / e, \qquad (125)$$

we need to calculate the rate at which $\varphi_1(x = 0, t)$ increases in the stationary limit in the presence of a potential drop $\sim -eV \int dx \ (\partial_x \varphi_+ / 2\sqrt{\pi}) \mathrm{sgn}(x)$, proportional to the applied voltage V (due to their equivalence, it does not matter which of the two modes we use for measuring the current). Eq. (125) reduces the transport problem to that of the dynamics of a quantum particle $q(t) \equiv \varphi_1(x = 0, t)$ subject to "damping" by all of the remaining bulk degrees of freedom, including those of the second mode. Therefore, we can employ methods of dissipative quantum mechanics [67,7] to solve this problem.

The effective action of the boundary mode S_{eff} is obtained by tracing over these remaining degrees of freedom

$$e^{-S_{eff}[q]} = \int \mathcal{D}[\varphi_1]\mathcal{D}[\varphi_2] \ \delta[q(\tau) - \varphi_1(x = 0, \tau)] \ \delta[\varphi_2(x = 0, \tau)] \ e^{-S}, \quad (126)$$

where

$$S = \int dx \int d\tau \ \left[\frac{1}{2v_F}(\partial_\tau \varphi_+)^2 + \frac{v_F}{2K_+^2}(\partial_x \varphi_+)^2 \right. \qquad (127)$$

$$\left. + \frac{1}{2v_F}(\partial_\tau \varphi_-)^2 + \frac{v_F}{2K_-^2}(\partial_x \varphi_-)^2 + \frac{4f_{bs}}{\pi\alpha^2}\varphi_-^2 \right], \qquad (128)$$

where $\pm$ combinations are defined in (44). Performing the integration, we get

$$S_{eff}[q] = \frac{1}{2} \int d\tau \int d\tau' \ q(\tau)\hat{\mathcal{K}}(\tau - \tau')q(\tau'), \qquad (129)$$

which accounts for the influence of the bulk modes exactly. In (126) we assume a large barrier so that incoherent single electron passages through subband 1 determine the current, while subband 2 is fixed by the barrier, cf. (126). Simultaneous contributions from subband 2 represent coherence effects, and are of higher order in the barrier transmission coefficient. The dynamics of q is governed by

$$S_0 = \int d\tau \left(W \cos 2\sqrt{\pi}q - \frac{eV}{\sqrt{2\pi}}q \right) + S_{eff}, \qquad (130)$$

which includes the applied voltage. Eq. (130) describes a damped quantum particle on a tilted washboard potential. Tunneling between adjacent minima of the potential corresponds to single electron transfers.

For large W, Eq. (130) maps onto a tight binding model with nearest neighbor hopping $\mathcal{T}$, whose value can in principle be deduced from W, assuming a δ-barrier : it is renormalized by damping compared to the bare value [68]. In this analogy, (125) corresponds to the particle's mobility, studied in Ref. [69]. In leading order $\sim |\mathcal{T}|^2$, the result is

$$I(V) = e|\mathcal{T}|^2 \int_0^\infty dt \; \sin(eVt) \, \text{Im} \, e^{-w(t)}, \tag{131}$$

where

$$w(t) = \int_0^\infty \frac{J(\omega)}{\omega^2} (1 - e^{-i\omega t}) \, . \tag{132}$$

The spectral function $J(\omega)$ is related to the Fourier transform $\mathcal{K}(\omega_n)$ of the kernel [67] appearing in (129),

$$J(\omega) = -\lim_{\eta \to 0} \text{Im} \, \mathcal{K}(-i\omega + \eta) = \frac{\omega}{K_+} + \frac{\sqrt{\omega^2 - \omega_0^2}}{K_-} \Theta(\omega - \omega_0), \tag{133}$$

with $\omega_0^2 = 8 f_{bs} v_F / \pi \alpha^2$. This yields

$$I(V) \sim V^{1/K_+ - 1}, \tag{134}$$

which gives $\alpha_d = 1/2K_+ - 1$ at voltages $V < \omega_0/e$. Above the gap, i.e, for $V \gg \omega_0/e$, gapless behavior of a two-subband Luttinger liquid is restored $I(V \gg \omega_0/e) \sim V^{1/K_+ + 1/K_- - 1}$.

The gapless behavior of fermion's DOS at the end of the wire can be interpreted in terms of a midgap state in the $(-)$ channel localized near the boundary. This is a bound state with zero energy formed in the potential well created by the static distortion of the φ_- field subject to an open boundary condition $\varphi_-(x = 0) = 0$, see Appendix D of Fabrizio and Gogolin paper [64]. Qualitatively, this is the same state which appears in a doped two-leg spin ladder, and which represents free $S = 1/2$ spin induced by a charged impurity, see, e.g., [70]. As a result, *local* density of states of the $(-)$ channel at the end of the wire takes the form $\rho_-(\epsilon) \sim \lambda \delta(\epsilon) + \rho_-^{\text{reg}}$, where ρ_-^{reg} stands for contribution of massive modes with energies *above* the CDW gap. On the other hand the $(+)$ mode remains gapless, and its end-chain DOS is given by $\rho_+(x = 0, \omega) \sim \omega^{\frac{1}{2K_+} - 1}$, [7,64,66]. The factor of 1/2 in the exponent is due to "factorization" of the electron operator into $(\pm)$ modes. Another consequence of such factorization is that DOS of a physical electron is a convolution of the DOS of the $(\pm)$ channels [cf. Eq. (116)]

$$\rho_{phys}(x = 0, \omega) \sim \int_0^\omega d\epsilon \rho_+(x = 0, \epsilon) \rho_-(x = 0, \omega - \epsilon). \tag{135}$$

Hence,

$$\rho_{phys}(x = 0, \omega) \sim \rho_+(x = 0, \omega) = \omega^{\frac{1}{2K_+} - 1}, \tag{136}$$

which implies $\alpha_d = 1/(2K_+) - 1$, in agreement with the result of the explicit calculation of $I(V)$ presented above, (134).

8   Experimental Consequencies and Conclusions

Apart from the rather well-known Luttinger-liquid phase, a two-subband quantum wire may also exhibit either a CDW or a (Cooper) superconducting phase. In both of these phases, certain modes of inter-subband charge- and spin-excitations are gapped, whereas the center-of-mass charge mode remains gapless. As a result, the conductance remains at the universal value of $2e^2/h$ per occupied subband, irrespective of whether the wire is in a gapless or gapped phase. However, the single-particle density of states in the middle of the wire has a hard gap. Above the gap, the DOS exhibits a non-universal threshold behavior $\rho(\omega) \sim \Theta(\omega - \Delta)(\omega - \Delta)^{-\beta}$, where $\beta \leq 1/2$. Softening of the threshold singularity is due "dressing" of the gapped modes by the remaining gapless one.

We find that the DOS for tunneling into the end of a wire in the CDW phase remains gapless, with the exponent determined by the center-of-mass mode only. This effect is due to frustration introduced into the CDW order by an open boundary (strong barrier).

Where should one look for such exotic phases of a quantum wire? We believe that quantum wires [11] prepared by cleaved edge overgrowth technique may be well-suited for observing the CDW phase. Indeed, the cross-section of such a wire is close to a square, which implies that the lowest states of transverse quantization should be close in energy. Hence, one-dimensional subbands can have close Fermi-momenta, which is a necessary condition for the formation of the CDW state. The Cooper phase, on the other hand, requires effective attraction in the relative charge channel, and has the best chance to occur when the second (upper) subband just opens for conduction. i.e., near the Van Hove singularity.

Acknowledgements

We would like to thank A. Abanov, B. Altshuler, M. Reizer, B. Marston, A. Nersesyan, T. M. Rice, S. Sachdev, and N. Shannon for interesting discussions and G. Martin for his help in manuscript preparation. DLM acknowledges the financial support from NSF DMR-970338 and Research Corporation (RI0082). WH acknowledges the kind hospitality of the University of Minnesota and support from the DFG (Germany) through contract HA 2108/4-1. The work at the University of Minnesota was supported by NSF Grants DMR-9731756 and DMR-9812340. DLM and LIG would like to thank the organizers of the workshop at Centro Stefano Franscini, Switzerland, where part of this work has been done, and the organizers of the WEH workshop in Hamburg.

References

1. H. J. Schulz, in *Proceedings of Les Houches Summer School LXI*, ed. E. Akkermans, G. Montambaux, J. Pichard, and J. Zinn-Justin (Elsevier, Amsterdam, 1995), p.533.
2. M. P. A. Fisher and L. I. Glazman in *Mesoscopic Electron Transport*, edited by L. P. Kouwenhoven, L. L. Sohn and G. Schön, Kluwer Academic, Boston, 1997.
3. M. Bockrath et al., Nature **397**, 598 (1999); cond-mat/9812233.
4. C. Dekker, Physics Today **52**, 22 (1999).
5. C. Ilani et al., cond-mat/9910116.
6. S. V. Zaitsev-Zotov et al., cond-mat/990756.
7. C. L. Kane and M. P. A. Fisher, Phys. Rev. B **46**, 15233 (1992).
8. K. A. Matveev and L. I. Glazman, Phys. Rev. Lett **70**, 990 (1993).
9. S. Tarucha, T. Honda, and T. Saku, Sol. St. Commun. **94**, 413 (1995).
10. K. J. Thomas et al., Phys. Rev. Lett **77**, 135 (1996).
11. A. Yacoby et al., Phys. Rev. Lett **77**, 4612 (1996); Solid State Communications **101**, 77 (1997).
12. A. M. Finkelstein and A. I. Larkin, Phys. Rev. B **47**, 10461 (1993).
13. M. Fabrizio, Phys. Rev. B **48**, 15838 (1993).
14. H. J. Schulz, Phys. Rev. B **53**, R2959 (1996).
15. L. Balents and M. P. A. Fisher, Phys. Rev. B **53**, 12133 (1996).
16. H. J. Schulz, cond-mat/9808167.
17. V. J. Emery, S. A. Kivelson, and O. Zachar, Phys. Rev. B **56**, 6120 (1997).
18. H. H. Lin, L. Balents, and M. P. A. Fisher, Phys. Rev. B **56**, 6569 (1997); cond-mat/9801285.
19. C. L. Kane, L. Balents and M. P. A. Fisher, Phys. Rev. Lett **79**, 5086 (1997).
20. Yu. A. Krotov, D. H. Lee. and S. G. Louie, Phys. Rev. Lett **78**, 4245 (1997).
21. R. Egger and A. O. Gogolin, Phys. Rev. Lett **79**, 5082 (1997).
22. R. Egger, A. O. Gogolin, Eur. Phys. J. B **3**, 281 (1998).
23. H. Yoshioka and A. Odintsov, Phys. Rev. Lett. **82**, 374 (1999).
24. Yu. A. Firsov, V. N. Prigodin, and Chr. Seidel, Phys. Rep. **126**, 245 (1985).
25. C. M. Varma and A. Zawadowski, Phys. Rev. B **32**, 7399 (1985).
26. J. Voit, Eur. Phys. J. B **5**, 505 (1999).
27. P. B. Wiegmann, Phys. Rev. B **59**, 15705 (1999).
28. E. Orignac and T. Giamarchi, Phys. Rev. B **56**, 7167 (1997).
29. M. Mori, M. Ogata and H. Fukuyama, J. Phys. Soc. Jpn. **66**, 3363 (1997).
30. O. A. Starykh and D. L. Maslov, Phys. Rev. Lett. **80**, 1694 (1998).
31. L. P. Kouwenhoven et al., Phys. Rev. Lett. **65**, 361 (1990).
32. K. B. Efetov and A. I. Larkin, Zh. Exp. Teor. Phys. **69**, 764 (1975) [Sov. Phys. -JETP **42**, 390 (1976)].
33. H. Frölich, J. Phys. C **1**, 544 (1968).
34. J. Ruvalds, Adv. Phys. **30**, 677 (1981).
35. F. D. M. Haldane, Phys. Rev. Lett. **45**, 1358 (1980).
36. T. Giamarchi, Phys. Rev. B **44**, 2905 (1991).
37. W. Metzner and C. Di Castro, Phys. Rev. B **47**, 16107 (1993).
38. G. D. Mahan, *Many-Particle Physics*, 2nd ed., Plenum Press, New York, 1990, section 4.4.
39. J. M. Luttinger, J. Math.. Phys. **4**, 1154 (1963).
40. V. J. Emery, *Highly Conducting One-Dimensional Solids*, eds. J. T. Devreese, R. E. Evrard and V. E. van Doren, New York, Plenum, p. 247 (1979).

41. S. Capponi, D. Poilblanc and T. Giamarchi, cond-mat/9909360.
42. K.A. Matveev, D. Yue, and L.I. Glazman, Phys. Rev. Lett. **71**, 3351 (1993).
43. V. L. Pokrovsky and A. L. Talapov, Sov. Phys. JETP **48**, 570 (1978).
44. H. J. Schulz, Phys. Rev. B **22**, 5274 (1980).
45. A. Luther, Phys. Rev. B **15**, 403 (1977).
46. T. Giamarchi and H. Schulz, Phys. Rev. B **39**, 4620 (1989)
47. J. B. Kogut, Rev. Mod. Phys. **51**, 701 (1979).
48. I. Safi and H. J. Schulz, Phys. Rev. B **52**, R17040 (1995).
49. D. L. Maslov and M. Stone, Phys. Rev. B **52**, R5539 (1995).
50. V. V. Ponomarenko, Phys. Rev. B **52**, R8666 (1995).
51. T. Giamarchi and H. Maurey, in *Correlated Fermions and Transport in Mesoscopic Systems*, edited by T. Martin, G. Montambaux, and J. Tran Thanh Van (Editions Frontieres, 1996), p. 13.
52. I. E. Dzyaloshinskii and A. I. Larkin, Sov. Phys. JETP **38**, 202 (1974).
53. A. M. Chang, L. N. Pfeiffer, and K. W. West, Phys. Rev. Lett. **77**, 2538 (1996).
54. A. V. Shytov, L. S. Levitov, and B. I. Halperin, Phys. Rev. Lett. **80**, 141 (1998).
55. A. Alekseev, V. Cheianov, A. P. Dmitriev, and V. Yu. Kachorovskii, cond-mat/9904076.
56. S. T. Chui and P. A. Lee, Phys. Rev. Lett. **35**, 315 (1975).
57. Z. Gulacsi and K. S. Bedell, Phys. Rev. Lett. **72**, 2765 (1994).
58. A. Luther and V. J. Emery, Phys. Rev. Lett. **33**, 589 (1974).
59. S. Sachdev, T. Senthil and R. Shankar, Phys. Rev. B **50**, 258 (1994).
60. D. G. Shelton, A. A. Nersesyan, and A. M. Tsvelik, Phys. Rev. B **53**, 8561 (1996).
61. A. M. Tsvelik, *Quantum Field Theory in Condensed Matter Physics*, Cambridge University Press, 1995.
62. R. Konik, F. Lesage, A. W. W. Ludwig, H. Saleur, cond-mat/9806334.
63. L. Balents, cond-mat/9902159.
64. M. Fabrizio and A. O. Gogolin, Phys. Rev. B **51**, 17827 (1995).
65. M. Fuentes, A. Lopez, E. Fradkin, and E. Moreno, Nucl. Phys. B **450**, 603 (1995).
66. A. O. Gogolin, A. A. Nersesyan and A. M. Tsvelik, *Bosonization and Strongly Correlated Systems*, Cambridge University Press, 1998.
67. U. Weiss, *Quantum Dissipative Systems*, Vol. 2 of *Modern Condensed Matter Physics*, World Scientific, Singapore, 1993; M. Sassetti and U. Weiss, Europhys. Lett. **27**, 311 (1994).
68. U. Weiss, Solid State Comm. **100**, 281 (1996).
69. U. Weiss et al., Z. Phys. B **84**, 471 (1991).
70. A. O. Gogolin, A. A. Nersesyan, A. M. Tsvelik, and Lu Yu, Nucl. Phys. B **540**, 705 (1999).

Interaction Effects in One-Dimensional Semiconductor Systems

Y. Tokura[1,2], A. A. Odintsov[1,3], and S. Tarucha[2,4]

[1] Department of Applied Physics, Delft University of Technology, Lorentzweg 1 2628 CJ Delft, The Netherlands
[2] NTT Basic Research Laboratories, Atsugi-shi, Kanagawa 243-0198, Japan
[3] Nuclear Physics Institute, Moscow State University, Moscow 119899 GSP, Russia
[4] University of Tokyo, 7-3-1 Hongo, Bunkyo-ku, Tokyo 113-0033, Japan

Abstract. The transport in a semiconductor quantum wire is discussed with recent experimental/theoretical results. The transport in a single-mode quantum wire shows a power-law dependence with a characteristic interaction parameter $K_w \sim 0.7$. We show temperature and bias dependences of the transport of interacting electrons through a single mode quantum wire subject to artificially modulating potential. The spin and charge separation is predicted to be observed in a periodically modulated system.

1 Introduction

One-dimensional (1D) interacting electron systems attract particular interests, since in the experiments, nearly ideal (long and clean) semiconductor quantum wire structures had been realized by virtue of recent developments in nanofabrication techniques. [1–3] More recently, a single wall metallic carbon nanotube begins to be manipulated in the electrical transport measurements. [4,5] From a theoretical point of view, an isolated 1D system is particularly interesting since it shows exotic non-Fermi liquid behavior, namely, the characteristics of a Tomonaga-Luttinger (TL) liquid. [6,7]

The characteristics of TL liquids, emerge in the power-law dependence of the physical quantities with excitation energy and the spin-charge separation in the excitation spectrum. A power-law dependence of the additional resistance on the temperature, T^{K_w-1}, with characteristic interaction parameter K_w, had been demonstrated in a nearly ballistic quantum wire in which a single 1D mode is occupied. [2] Similar behaviors and power-law dependences on the applied bias were observed recently in a single wall metallic carbon nanotube. [5] Although the system of the carbon nanotubes is ideal for studying TL liquid behavior (besides its complication of the occupation of two channels), there are many good reasons to study the semiconductor quantum wire system. We believe the biggest one is that its Fermi wave length, compared with that of the carbon nanotubes (~ 0.1nm), is rather long (~ 10nm), where the current nanofabrication technique can taylor the potential in this length scale. In the second part of this paper, we demonstrate several examples related to this point. The progress in the experiments of the semiconductor wire is not so rapid, while there are a number of efforts accumulating. We would like to report some recent results on

1D transport and critically discuss the potential problems and new physics in this system.

The article is made up of two parts. In Sec. 2, we discuss the experimental aspects of the recent research on semiconductor quantum wires. Section 3 is devoted to the discussion of the transport properties in a modulated quantum wire. Finally we concludes in Sec. 4.

2   Semiconductor Wires

2.1   Ballistic Transport

In order to see the interacting 1D behavior in a quantum wire, we need stronger lateral confinement potential to avoid multiple-channel transport for finite temperatures or for finite bias conditions. Another important criterion is the cleanness of the system. The quantization of the transverse momentum in a ballistic two-dimensional electron gas end ups in a conductance quantization in units of $G_0 = 2e^2/h$. The transport of a finite quantum wire of length L obeys Landauer's formula if the interaction effect is absent. The conductance quantization is predicted only if the effect of backscatterings in the channel is negligible AND the contacts are ideal, namely, the incoming current flows are only governed by the contact (quasi-equilibrium) chemical potentials and outgoing current flows are completely absorbed in the contacts. The wires formed by cleaved-edge-overgrowth method had shown a strange contact resistance although the transport in the wire seems ballistic. [3] Similar effect of a contact resistance had been found in V-grooved quantum wires. [8] For these experiments, a mechanism of the contact resistance, backscatterings in the contacts, is proposed, [9] although further studies seem necessary to clear the issue, for example, the magnetic field dependences of the contact resistance. [10] For relatively short quantum wires (short constrictions of $L \sim 1\mu m$), there have been several reports on an anomalous conductance structure at $\sim 0.7G_0$. [11] As for the origin of this structure, it is controversial whether it comes from the interaction effect IN the constrictions.

Now in general, if the interaction is switched on in the wire, how is the conductance renormalized ? At least for a homogeneous and long wire attached to large contacts, the experiments had shown that the conductance is quantized with G_0 if there is negligible backscattering in the channel. [2] Soon after the experiments, the physical reason of this absence of conductance renormalization by the interaction was presented theoretically. [12–14] If there is some source of backscatterings, the wire conductance is drastically changed from that of the noninteracting one.

2.2   Power-Law Dependence

In this subsection, we describe a detail of our system and discuss the experimental results.

The wire structures are formed in $Al_{0.35}Ga_{0.65}As/GaAs$ modulation-doped heterostructures. The distance between the heterostructure and the top surface

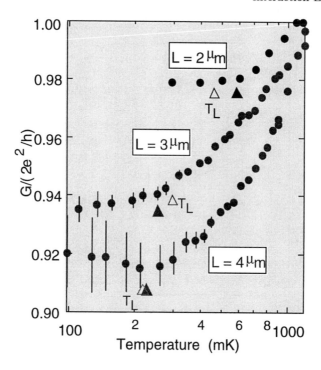

Fig. 1. Temperature vs. the conductance of wires with three different lengths. Solid triangles show the saturation points of the temperature dependence. Open triangles are T_L calculated from the wire length L.

are very large (680nm) and by virtue of very thick (75nm) undoped spacer layer, the effect of remote ionized impurity scattering is very small. The electron mobility and the electron density of the original two-dimensional sample at 1.5 K are $7.8 \times 10^6 \text{cm}^2/\text{Vs}$ and $2.3 \times 10^{11}\text{cm}^{-2}$, respectively, after illumination. This corresponds to the mean free path l of about 60μm. Successive electron beam lithography and a lift-off process define a Ti line mask, and BCl_3 plasma etching with a subsequent wet etch makes the wire structure of about 0.6μm width at the heterointerface. In addition to that, Schottky gates are selectively formed on the chemically etched slopes adjacent to the wire, which help to form a strong lateral confinement potential of the wire.

The conductance is measured for various temperature T from 0.29 to 1.2 K using standard lock-in technique. In Fig. 1, we plotted the wire conductance as a function of the temperature for three wires with different length. The gate voltage is chosen so that the conductance is in the most flat region of the first plateau at $T \sim 1$K. Two important features can be seen. The first is that the conductance decreases with T from the value of the quantized conductance. The second is that the decrease eventually saturates and for lower temperatures the conductance falls in a more or less constant value depending on the length of the wire, L. The characteristic temperature T_L where the conductance decrease

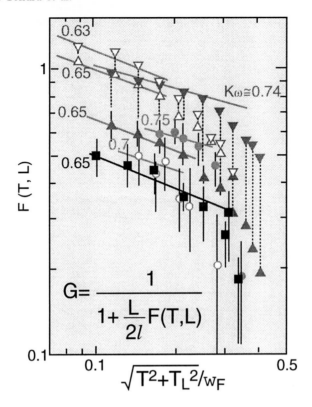

Fig. 2. $F(T,L)$ vs. renormalized temperature evaluated for various wires: the circles $L = 2\mu m$, the triangles $L = 5\mu m$, and the squares $L = 10\mu m$. The filled and open symbols are for dark and after illumination conditions, respectively.

stops is a monotonically decreasing function of L. These features are clearly the evidence of the interaction effects in the quantum wire, since a two-dimensional electron gas possesses almost no temperature dependent scattering mechanisms at such a low temperature range. TL-liquid theory predicts a definite temperature dependence of the conductance in a system with a repulsive interaction ($K_w < 1$) and finite scattering potential (irrespective to the strength). [7] However, for a finite wire connected to large contacts, this TL-liquid behavior cannot persist until absolute zero since the physics of the low energy scale is successively controlled by the Fermi liquid property of the contacts. The theory predicts such a transition occurs at $k_B T_L \sim \hbar v_F/L$ where v_F is the Fermi velocity and k_B is the Boltzmann constant. [15,16] The results shown in Fig. 1 are consistent with the prediction. Ogata and Fukuyama have proposed a universal function $F(T,L) \sim (\sqrt{T^2 + T_L^2}/\omega_F)^{K_w-1}$ where $\hbar\omega_F$ is the Fermi energy. [15] The conductance is given by $G_0/(1 + F(T,L)L/2l)$ where l is the mean free path. By fitting the data with the function as shown in Fig. 2, the value of the interaction parameter K_w is calculated to 0.63 to 0.75. [2]

The power exponent is not a universal value and depends on the microscopic properties of the scattering mechanism. We found that the power exponent is smaller for higher conductance plateau ($G \sim nG_0$). [17] This result is easily understood since for higher plateau, the electron density is larger and the Coulomb interaction is effectively screened. Therefore, the interaction parameter K_w, which is roughly estimated by $(1 + 2U_0/(\pi v_F))^{-1/2}$ with a characteristic Coulomb energy U_0, approaches to 1 with n. Microscopic calculations are found in Ref. [18–20].

In the TL liquid, the current vs. bias is also predicted to show power-law behavior. We also observed a non-linear characteristics if the bias exceeds a characteristics value determined by the wire length. [17] To analyze the results, we may need to consider the effect of the transport in the higher subbands.

3 Modulated Wires

In semiconductor quantum wires, the potential profile can be effectively controlled by geometrical or electrostatic confinement which can be achieved by etching the heterostructure or depositing metallic gates, respectively. In this section, we discuss the effect of the double barrier potential and the periodic potential on a 1D electron transport within a perturbation approach.

We use the following phase Hamiltonian:

$$H_\rho = \int \frac{dx}{\pi} \hbar v_\rho(x) [\frac{1}{K_\rho(x)} (\nabla \theta_\rho(x))^2 + K_\rho(x)(\nabla \phi_\rho(x))^2], \qquad (1)$$

where $\theta_\rho(x)$ and $\phi_\rho(x)$ are phase operators for the charge sector, satisfying the commutation relation

$$[\theta_\rho(x), \phi_\rho(x')] = \frac{\pi i}{4} \text{Sign}(x - x'), \qquad (2)$$

$K_\rho(x)$ being the interaction parameter and $v_\rho(x)$ being the velocity of charge excitation. The effect of a slowly varying part of the potential along the wire can be incorporated into the inhomogeneous 1D model where $K_\rho(x)$ and $v_\rho(x)$ are position dependent. We assume constant repulsive interaction, $K_\rho = K_w < 1, v_\rho = v_w = v_F/K_w$, in the 1D region, $| x | < L/2$, and non-interacting electrons in two-dimensional contacts $K_\rho(x) = 1$. [12–14] The components of the charge density and current slowly varying on the scale of k_F^{-1} are given by $n(x) = -\frac{2e}{\pi}\nabla\theta_\rho(x)$ and $j(x) = \frac{2e}{\pi}\frac{\partial\theta_\rho(x)}{\partial t}$. A similar Hamiltonian with phase operators θ_σ and ϕ_σ and $K_\sigma = 1$ describes the spin sector. It is important to notice that the spin and charge degrees of freedom are decoupled in the Hamiltonian, $H_0 = H_\rho + H_\sigma$

The Fermi operators for the right ($r = +1$) and left ($r = -1$) moving electrons with spin s are given by

$$\Psi_{rs}^\dagger(x) = \sqrt{\rho}e^{ir(k_F x + \theta_s) + i\phi_s} \qquad (3)$$

where $\theta_s = \theta_\rho + s\theta_\sigma$ and $\phi_s = \phi_\rho + s\phi_\sigma$. The constant $\rho = (2\pi\alpha)^{-1}$ is related to a cut-off length α, which is of the order of the Fermi wave length.

3.1 Born Approximation

We introduce a potential for electron backscattering, $V(x)$, finite only in the wire region, and weak enough that we can treat it perturbatively, $\mid V \mid \ll E_F$, where E_F is the Fermi energy. Using the field operators, the perturbing term is

$$H_v = 4\rho \int dx V(x) \cos(2k_F x + 2\theta_\rho(x)) \cos(2\theta_\sigma(x)), \tag{4}$$

and now the spin and charge sector are coupled by the potential. The ϕ operators do not emerge in H_v.

The current through the quantum wire corresponds to the shift of the Fermi point for right- and left-moving electrons away from their equilibrium position $k_{FR(L)} \rightarrow k_{FR(L)} \pm \frac{eA}{\hbar c}$, where A is a constant vector potential. This can be described by adding the phase factor to the bosonic field, $\phi_s \rightarrow \phi_s + \frac{ie}{\hbar c} \int^x A dx'$, see Eq. (3). The gauge transformation results in the Hamiltonian, Eq. (1)

$$H_\rho^A = \int \frac{dx}{\pi} \hbar v_\rho(x) [\frac{1}{K_\rho(x)} (\nabla \theta_\rho(x))^2 + K_\rho(x)(\nabla \phi_\rho(x) - \frac{eA}{\hbar c})^2].$$

Time evolution of the field operator θ_ρ is given by Heisenberg equation. We decompose the field $\theta_\rho(x,t)$ like $\theta_{cl}(t) + \tilde{\theta}_\rho(x,t)$. In the absence of backscattering in the wire (the Hamiltonian is now $H_{0A} = H_\rho^A + H_\sigma$), the classical part is given by $\theta_{cl}(t) = \frac{\Omega}{2}t$, with $\Omega = \frac{2ev_F}{\hbar c}A$, where the relation $v_\rho(x)K_\rho(x) = v_F$ is taken into account. This solution describes the current $j = \frac{e\Omega}{\pi}$ flowing through the wire.

The shift of the Fermi points $\frac{eA}{\hbar c}$ for the right- and left-moving electrons corresponds to the difference of the chemical potentials $\mu_L - \mu_R = \hbar\Omega$ of the left and right contacts. This agrees with Landauer's formula $ej/(\mu_L - \mu_R) = G_0$. The electrochemical potential, $\mu(x)$, is obtained by $-(\hbar v_\rho(x)/K_\rho(x)) \langle \nabla \theta_\rho(x,t) \rangle_H$, where the average is evaluated over the total Hamiltonians $H = H_{0A} + H_v$. In the presence of the electron backscattering, the variation of the electrochemical potential along the wire is given by

$$\nabla \mu'(x) = \frac{1}{i} \langle \nabla[H_v, \phi(x,t)] \rangle_H$$
$$= 4\pi \rho V(x) \langle \cos 2\theta_\sigma(x,t) \sin(2k_F x + 2\tilde{\theta}_\rho(x,t)) \rangle_H. \tag{5}$$

The additional increase of the voltage difference between left and right contacts is obtained by integrating Eq. (5) over the wire region and estimating the average (5) in the lowest order in $V(x)$,

$$\delta\mu = \mu'(-\frac{L}{2}) - \mu'(\frac{L}{2}) \tag{6}$$

$$= \frac{\pi i}{\hbar}(4\rho)^2 \int_{-L/2}^{L/2} dx V(x) \int_{-L/2}^{L/2} dx' V(x') \int_{-\infty}^{t} dt'$$

$$\times \langle [\cos(2\theta_\sigma) \sin(2k_F x + 2\tilde{\theta}_\rho), \cos(2\theta_\sigma') \cos(2k_F x' + 2\tilde{\theta}_\rho')] \rangle_{H_{0A}}, \tag{7}$$

where $\tilde{\theta}_\rho \equiv \tilde{\theta}_\rho(x,t), \tilde{\theta}'_\rho \equiv \tilde{\theta}_\rho(x',t')$ etc. for compact notations. For a linear transport regime, the increase of the resistance is expressed by $\Delta R = R_0 \delta\mu/(\hbar\Omega)$ where $R_0 = 1/G_0$. For nonlinear regime, $\delta\mu$ is related to the backscattering current $I_{bs} \equiv G_0 \delta\mu/e$.

The commutator is evaluated to

$$\frac{i}{2}\sin(2k_F(x-x')+\Omega(t-t'))\mathrm{Im}[\langle e^{2i\tilde{\theta}_\rho(x,t)}e^{-2i\tilde{\theta}_\rho(x',t')}\rangle_{H_\rho}\langle e^{2i\theta_\sigma(x,t)}e^{-2i\theta_\sigma(x',t')}\rangle_{H_\sigma}]$$

and

$$\langle e^{2i\tilde{\theta}_\rho(x,\tau)}e^{-2i\tilde{\theta}_\rho(x',0)}\rangle_{H_\rho} \equiv \exp(C_\rho(x,x',\tau))$$

where $\tau = t - t'$ since H_ρ is homogeneous in time. The correlator is

$$C_\rho(x,x',\tau) = \tag{8}$$
$$= \int \frac{\hbar d\omega}{\pi}(1 + \coth\frac{\hbar\omega}{2k_BT})e^{-\frac{\omega}{\omega_c}}\mathrm{Im}[G_\omega^R(x,x) + G_\omega^R(x',x') - 2e^{-i\omega\tau}G_\omega^R(x,x')].$$

We introduced frequency cutoff ω_c of the order of $E_F/\hbar$ to avoid divergence. The retarded Green's function is given by

$$\mathrm{Im}G_\omega^R(x,x') = -\frac{\pi K_w(1-R)}{4\hbar\omega}\frac{(1+R^2)\cos q(x-x') + 2R\cos qL\cos q(x+x')}{1 - 2R^2\cos 2qL + R^4},\tag{9}$$

for $|x|, |x'| < L/2$ and $R = (1 - K_w)/(1 + K_w), q = \omega/v_w$. In the low frequency limit $(\omega \ll v_w/L)$, $\mathrm{Im}G_\omega^R(x,x') \sim -(\pi/4\hbar\omega)\cos q(x-x')$. For high frequencies, $(v_w/(L-|x+x'|) \ll \omega)$, one can average over rapidly oscillating component with the result $\mathrm{Im}G_\omega^R(x,x') \sim -(\pi K_w/4\hbar\omega)\cos q(x-x')$. For the intermediate frequency range, $(v_w/L \ll \omega \ll v_w/(L-|x+x'|))$, $\mathrm{Im}G_\omega^R(x,x') \sim -(\pi K_w/4\hbar\omega)[\cos q(x-x') + R\cos q(L-|x+x'|)]$ since the boundary effect near the contacts becomes important. [24] Finally, for noninteracting electrons $(K_w = 1)$, the correlator C is given by

$$C_\sigma(x,x',\tau) = \frac{1}{2}\sum_\pm \ln[\frac{\pi\tau_\pm k_BT/\hbar}{\sinh(\pi\tau_\pm k_BT/\hbar)}\frac{1}{1 + i\omega_c\tau_\pm}],\tag{10}$$

where $\tau_\pm = \tau \pm (x-x')/v_F$.

3.2   Double Barrier Scattering

Now we apply Eq. (6) for the case of two short range impurities located symmetrically in the wire, $V(x) = V_0[\delta(x+\frac{b}{2})+\delta(x-\frac{b}{2})]$. The problem has been already considered for an infinite system in weak and strong scattering limits.[25–27] The backscattering current is given by

$$I_{bs} = -\frac{\pi}{e\hbar}G_0U_{imp}^2\int_0^\infty d\tau\sin(\Omega\tau)\mathrm{Im}[\exp(C_\rho(\frac{b}{2},\frac{b}{2},\tau) + C_\sigma(\frac{b}{2},\frac{b}{2},\tau))$$
$$+ \cos(\xi b)\;\exp(C_\rho(\frac{b}{2},-\frac{b}{2},\tau) + C_\sigma(\frac{b}{2},-\frac{b}{2},\tau))],\tag{11}$$

where $\xi = 2(k_F - \pi/2b)$ and $U_{imp} = 4\rho V_0$. The term related to the pair $(b/2, b/2)$, I_{bs}^{IC}, corresponds to the incoherent scatterings from the potential at $x = \pm b/2$. The coherent term related to the pair $(b/2, -b/2)$, I_{bs}^{C}, describes the electron interference due to the scattering from two impurities. It oscillates depending on the Fermi energy due to the factor, $\cos(\xi b)$.

For noninteracting electrons, $I_{bs}^{IC} = I_0(\pi U_{imp}/\hbar\omega_c)^2/2$, which is just an Ohmic resistance independent of the bias and the temperature, where $I_0 = G_0\hbar\Omega/e$. The coherent term,

$$I_{bs}^{C} = I_{bs}^{IC} \cos(\xi b) \frac{\frac{2\pi b k_B T}{\hbar v_F}}{\sinh \frac{2\pi b k_B T}{\hbar v_F}} \frac{\sin \frac{b\Omega}{v_F}}{\frac{b\Omega}{v_F}}, \tag{12}$$

decays exponentially with temperature and oscillates as a function of applied bias.

We then consider the case of finite repulsive interaction, $K_w < 1$. At high temperatures, $\hbar v_w/(L - b) \ll k_B T$, the incoherent part in the linear response regime ($\Omega \ll v_w/L$) is

$$I_{bs}^{IC} \sim I_0(\frac{\pi}{\hbar\omega_c})^{1+K_w} U_{imp}^2 (k_B T)^{K_w-1}. \tag{13}$$

The power-law dependence T^{K_w-1} has also been found for the cases of single impurity and many impurity scattering. [7,15,16] The coherent part decays with temperatures faster than in noninteracting case Eq. (12),

$$I_{bs}^{C} \sim I_{bs}^{IC} \cos(\xi b) \frac{b k_B T}{\hbar v_w} e^{-(K_w + \frac{1}{K_w})\frac{\pi b k_B T}{\hbar v_w}}. \tag{14}$$

The formula for the intermediate temperature regime $\hbar v_w/L \ll k_B T \ll \hbar v_w/(L-b)$, can be obtained from Eqs. (13,14), by replacing K_w by $2K_w/(1 + K_w)$. Figure 3 is the temperature dependence calculated numerically.

For low temperatures and linear transport regime, the result is the same as the non-interacting case. For specific values of the Fermi energy where $\cos(\xi b) = -1$ is satisfied, the perturbation considered here gives zero correction. For large bias, $v_w/(L-b) \ll \Omega$, the incoherent term is $I_{bs}^{IC} \sim U_{imp}^2(\hbar\Omega)^{K_w}(\hbar\omega_c)^{-K_w-1}G_0/e$ which gives sub-linear characteristics. For the intermediate regime, $v_w/L \ll \Omega \ll v_w/(L - b)$, the bias dependence is $\Omega^{2K_w/(K_w+1)}$. The coherent term is now

$$I_{bs}^{C} \sim I_{bs}^{IC} \cos(\xi b)[A_\rho(\frac{b\Omega}{v_w})^{-1-\frac{K_w}{2}} \cos(\frac{b\Omega}{v_w} - \frac{\pi K_w}{4})$$

$$+ A_\sigma(\frac{b\Omega}{v_w})^{-\frac{1}{2}-K_w} \cos(\frac{b\Omega}{v_F} - \frac{\pi(2K_w + 1)}{4})], \tag{15}$$

where A_ρ and A_σ are constants of order unity. We show the numerical results in Fig. 4. The oscillation with Ω has two periods, one from the charge velocity $2\pi v_w/b$ and the other from the spin velocity $2\pi v_F/b$. The effects of the spin-charge separation have been discussed in the context of the local density of states near the open boundary of quantum wire,[29] and the transport in open quantum dots.[30]

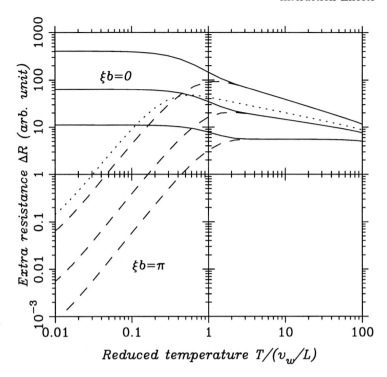

Fig. 3. Temperature dependence of the extra resistance ΔR of a wire with double barrier potential. The solid lines ($\xi = 0$) and the dashed lines ($\xi b = \pi$) are from top to bottom, $K_w = 0.5, 0.75, 1.0$ and $b = 0.4L$. For $k_B T \gg \hbar v_w/L$, they shows power-law dependence like T^{K_w-1}. For lower temperature, T^2 for $\xi b = \pi$ otherwise constant. The dotted line is $K_w = 0.5$ and $\xi b = \pi$ with $b = L$, which shows different temperature power $T^{(K_w-1)/(K_w+1)}$ for higher temperatures.

3.3 Periodic Potential Scattering

We investigate the transport of interacting electrons through a single mode quantum wire subjected to a periodic electrostatic potential due to modulation of the donor-layer widths (Fig. 5).

The potential in the wire region is given by $V(x) = W_0 \cos(Gx)\Theta(\frac{L}{2} - |x|)$, where $\Theta(x)$ is the step function and $G = 2\pi/a$, with the period of the potential a. We will consider weak periodic potential, $W_0 \ll E_F$, which can be treated perturbatively. We will concentrate on the single-particle Bragg reflection near the full-filling condition, $2k_F = G$.

The backscattering current is given by [28,32]

$$I_{bs} = -\frac{2\pi}{e\hbar}(W_0\rho)^2 G_0 \times$$

$$\times \int_0^\infty d\tau \int dx \int dx' \sin(\xi(x - x') + \Omega\tau)\operatorname{Im} e^{C_\rho(x,x',\tau)+C_\sigma(x,x',\tau)}, (16)$$

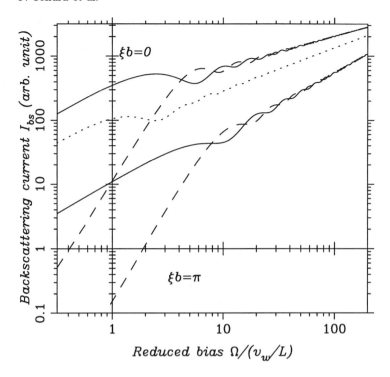

Fig. 4. Bias dependence of the backscattering current I_{bs} in a wire of double barrier potential. The solid lines ($\xi = 0$) and the dashed lines ($\xi b = \pi$) are from top to bottom, $K_w = 0.5, 1.0$ and $b = 0.4L$. For $\Omega \gg v_w/L$, they shows power-law dependence like Ω^{K_w}. For lower bias, Ω^3 for $\xi b = \pi$ otherwise Ω linear. The oscillating component is periodic with Ω and made up of two periods (for $K_w = 0.5$) from charge and spin sector, although it is hard to see in the log-log plot. The dotted line is $K_w = 0.5$ and $\xi b = 0$ with $b = L$, which shows different bias power $\Omega^{2K_w/(K_w+1)}$.

where $\xi = 2k_F - G$.

Non-interacting electrons First we consider the limiting case of non-interacting electrons ($K_w = 1$). In the linear transport regime at high temperatures, $\hbar v_F/L \ll k_B T \ll \hbar v_F \xi$, we obtain from Eq. (16),

$$\Delta R = 2\Xi_1 R_0 \left(\frac{\pi \hbar v_F}{8 L k_B T} \frac{1}{\cosh^2 \frac{\hbar v_F \xi}{4 k_B T}} + \frac{1}{(L\xi)^2} \right), \tag{17}$$

with $\Xi_1 \equiv (W_0 \rho L)^2 (\frac{\pi}{\hbar \omega_c})^2$. At even higher temperatures, $k_B T \gg \hbar v_F/L, \hbar v_F \xi$, the second term in Eq. (17) should be disregarded. Therefore at full-filling, ΔR is inversely proportional to the temperature. Away from the full-filling, the exponential suppression of the electron backscattering with decreasing temperature crosses over to temperature independent regime as shown in Fig. 6a.

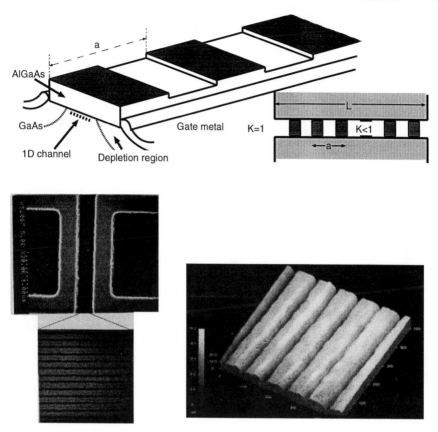

Fig. 5. Upper:Schematic diagram of the periodically modulated 1D system. The periodic potential is made by controlling the distance of the 1D channel and the surface. The density of the wire is controlled by the gate from the sides. Lower left:Top view of the device by scanning electron microgr̈aph. Bright regions are gate metals with gap of $0.4\mu m$. Lower is the magnified view where the periodic surface modulation is visible with the period of $60nm$. Lower Right:Surface topography of the corrugated surface by atomic force micrograph. The period is about $60nm$ and the depth is about $10nm$.

In the non-linear transport regime at $T = 0$, I_{bs} has the form, $\Xi_1 \frac{\hbar v_F}{L} \times$
$\times \sum_{\pm} (\text{Si}(\kappa_{\pm} L) + \cos(\kappa_{\pm} L - 1)/(\kappa_{\pm} L))G_0/e$, where $\kappa_{\pm} = \frac{\Omega}{v_F} \pm \xi$ and $\text{Si}(x)$ is the sine integral. For the full-filling condition, $\xi = 0$, I_{bs} increases linearly with Ω, $I_{bs} \sim 2\Xi_1 I_0(1 - \cos(\xi L))/(\xi L)^2$ at low bias ($\Omega \ll v_F/L$) and saturates to a constant $\Xi_1 \pi(\hbar v_F/L)G_0/e$ at high bias. Away from full-filling, the dependence of $I_{bs}(\Omega)$ shows threshold singularities at $\Omega \sim v_F \xi$ with a finite width of v_F/L. [33] The numerical results are shown in Fig. 8

These results are equivalent to that derived from the Landauer type approach. The latter expresses the extra resistance ΔR due to the periodic poten-

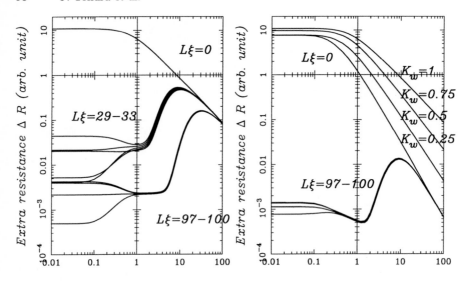

Fig. 6. Temperature dependence of extra resistance ΔR. a (left): noninteracting case $K_w = 1$ for several values of $L\xi = 0, 29 - 33, 97 - 100$ with step 1. b (right): Interaction dependence for $L\xi = 0$ with $K_w = 1, 0.75, 0.5, 0.25$. We also plot for $L\xi = 97 - 100$ with $K_w = 0.25$ which shows extra power-law temperature dependence for $T \sim T_L$.

tial,

$$\Delta R = R - R_0 = R_0 \langle |r(k)|^2 \rangle_T, \tag{18}$$

via the reflection coefficient $r(k)$ for an incoming electron with wave number k ($\langle \cdots \rangle_T$ denotes the thermal average and $\langle |r(k)|^2 \rangle_T \ll 1$ is assumed). In the Born approximation, $r(k) = \gamma \frac{\sin \xi_k L/2}{\xi_k L/2}$, with $\xi_k = 2k - G$, and $\gamma = k_F L W_0/(2E_F)$. At low temperatures, $k_B T \ll \frac{\hbar v_F}{L}$, $\langle |r(k)|^2 \rangle_T = |r(k_F)|^2$ has a sharp maximum γ^2 at full-filling and ΔR is independent of T.

At high temperatures $k_B T \gg \frac{\hbar v_F}{L}$,

$$\langle |r(k)|^2 \rangle_T = \gamma^2 \left[\frac{\pi \hbar v_F}{8 L k_B T} \cosh^{-2} \frac{\hbar v_F \xi}{4 k_B T} + \frac{1}{(L\xi)^2} \right], \tag{19}$$

with $\xi \equiv \xi_{k_F}$. Equation (19) is equivalent to Eq. (17) with replacing γ^2 with $2\Xi_1$, (the latter quantities are equal if $\alpha = 1/k_F$ and $\hbar \omega_c = 2E_F$).

Similarly, the backscattering current in the non-linear response regime at zero temperature is given by $I_{bs} = I_0 \langle |r(k)|^2 \rangle_V$, where $\langle \cdots \rangle_V$ is the average over the energy range of $[\mu_R, \mu_L]$. Again, the obtained nonlinear characteristics are identical to the one shown previously.

Interacting electrons We will now consider the system with finite interaction. We first discuss spinless system, which is described by Eq. (16) with

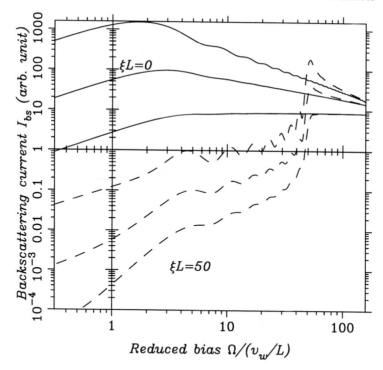

Fig. 7. Bias dependence of backscattering current I_{bs} at $T = 0$ in a wire of spinless electrons with periodic potential. The solid lines ($\xi = 0$) and the dashed lines ($\xi L = 50$) are from top to bottom, $K_w = 0.5, 0.75, 1.0$. For high bias, $\Omega \gg v_w/L$, they show power-law dependence like $\Omega^{2(K_w-1)}$. For lower bias, they show Ω linear dependence. The singular threshold characteristics are found for $\Omega = 50 v_w/L$ for $\xi L = 50$ curves.

$C_\sigma = 0$, $C_\rho \to 2C_\rho$ and $W_0 \to W_0/2$. The spinless model could be applicable to spin polarized electrons in a strong magnetic field. In the linear transport regime and at high temperature, $k_B T \gg \frac{\hbar v_w}{L}, \hbar v_w \xi$, we obtain $\Delta R \sim R_0 \Xi_{K_w}(\hbar v_w/L)(k_B T)^{2K_w-3}$. We used an interaction dependent factor $\Xi_{K_w} \equiv (W_0 \rho L)^2 (\frac{\pi}{\hbar \omega_c})^{2K_w}$. In the intermediate temperature range, $\hbar v_w/L \ll k_B T \ll \hbar v_w \xi$, the temperature dependence of the resistance crosses over from exponential suppression,

$$\Delta R \sim R_0 \Xi_{K_w}(\hbar v_w \xi/\pi)^{2K_w-2} \frac{\hbar v_w}{L k_B T} e^{-\hbar v_w \xi/(2k_B T)}, \qquad (20)$$

for $k_B T^* \ll k_B T \ll \hbar v_w \xi$, to power-law decay,

$$\Delta R \sim R_0 \Xi_{(1+R)K_w}(\frac{\omega_c}{v_w \xi})^{2RK_w} \frac{1}{\xi^2}(k_B T)^{2(K_w-1)/(K_w+1)}, \qquad (21)$$

for $\hbar v_w/L \ll k_B T \ll k_B T^*$. T^* is defined by the temperature when ΔR's of Eqs.(20,21) coincide. At even lower temperatures, $k_B T \ll \hbar v_w/L$, the resistance

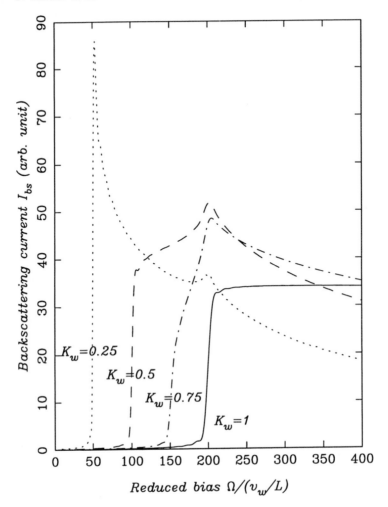

Fig. 8. Bias dependence of the backscattering current I_{bs} of a wire of spinfull electrons with periodic potential at $T = 0$. The curves are for $K_w = 1, 0.75, 0.5, 0.25$ and $\xi L = 200$. The bias is normalized by $\hbar v_w/L$, ant the lower threshold occurs at $K_w \hbar v_w/L = \hbar v_F/L$.

shows characteristic oscillations as a function of filling factor ξ. Away from commensurable filling, the exponential suppression of ΔR with decreasing energy E is universal, however, there is power-law suppression in its cross-over to the temperature independent limit.

In non-linear transport regime, $v_w/L \ll \Omega$, at zero temperature, we have
$$I_{bs} \sim \Xi_{K_w}(\hbar v_w/L)(\hbar\Omega)^{2(K_w-1)}[(1+\tfrac{v_w\xi}{\Omega})^{K_w-1}+(1-\tfrac{v_w\xi}{\Omega})^{K_w-1}\Theta(\Omega-v_w\xi)]G_0/e.$$
The current shows power-law behavior $I_{bs} \propto \Omega^{2(K_w-1)}$ for $\Omega \gg v_w\xi$ and has a singularity $(\Omega - \Omega_c)^{K_w-1}$ for $\Omega \sim v_w\xi$. The numerical result is shown in

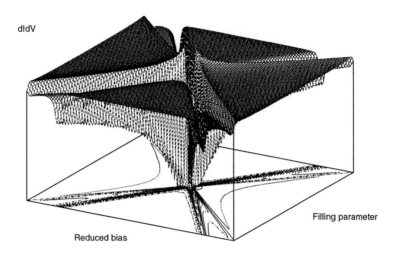

dIdV

Filling parameter

Reduced bias

Fig. 9. Differential conductance dI/dV of a wire of spinfull electrons with periodic potential at $T = 0.5\hbar v_w/L$ and $K_w = 0.7$. The bias Ω is between $\pm 100 v_w/L$ and the filling $L\xi$ is between ± 100. The zero of dI/dV locates at the flat region of the surface plot. Four lines of singularity at $\Omega/(v_w/L) = \pm L\xi$ (for charge sector) and $\Omega/(v_F/L) = \pm L\xi$ (for spin sector) are clearly seen.

Fig. 7. The oscillation with a period $2\pi v_w/L$ in Fig. 7.is related to a resonant backscattering process with excitation of virtual plasmon modes.[31]

The extension to the spinfull case is straightforward. In particular the exponents (with exception of Eq. (21)) can be obtained by replacing K_w with $(K_w + 1)/2$, i.e., for high temperatures and linear regime $\Delta R \propto (k_B T)^{K_w - 2}$ and the intermediate anomalous regime $\Delta R \propto (k_B T)^{(K_w - 1)/(K_w + 1)}$. The numerical results for the temperature dependence of ΔR are shown in Fig. 6b. For high bias regime at $T = 0$, $I_{bs} \propto (\hbar\Omega)^{K_w - 1}$. Nonlinear I-V characteristics (Fig. 8) show additional feature, namely, two-step threshold singularities at $\Omega = v_F \xi$ and $\Omega = \hbar v_w \xi$,

$$I_{bs} \sim \Xi_{\frac{K_w + 1}{2}} \frac{\hbar v_w}{L} (\hbar\Omega)^{K_w - 1} [B_\rho ((1 + \frac{v_w \xi}{\Omega})^{K_w - 1} + (1 - \frac{v_w \xi}{\Omega})^{K_w - 1} \Theta(\Omega - v_w \xi))$$

$$+ B_\sigma ((1 + \frac{v_F \xi}{\Omega})^{K_w - 1} + (1 - \frac{v_F \xi}{\Omega})^{K_w - 1} \Theta(\Omega - v_F \xi))] G_0/e, \qquad (22)$$

where $\xi L \mid v_w - v_F \mid /v_F \gg 1$, B_ρ and B_σ are constants of order unity. The two-step threshold singularities can be best seen in the plot of differential conductance as a function of filling factor ξL and applied bias Ω, Fig. 9. The experimental observation of such singularities would confirm the concept of *spin-charge separation* in 1D interacting system.

Finally, we discuss the feasibility of fabricating modulated quantum wires. Although the effect of impurity scattering can be distinguished since there is a difference between the power-law dependence caused by impurity scattering and that caused by Bragg reflection, it is preferable to use a clean wire to gain larger signal from the Bragg reflection. We chose an AlGaAs/GaAs heterostructure with very high mobility as the starting material. However, we could not take advantage of the high mobility since the required carrier density is too high to allow us to realize commensurate periodic potential with the current technique ($a \sim 40$ nm). Therefore, we have to reduce the carrier density to the order of $1 \times 10^{11} cm^{-2}$, and ballistic transport is expected only for an L of less than several microns, which gives $k_F L \sim 100$. In our perturbative approach ΔR or I_{bs} is totally proportional to γ^2. To obtain a reasonable value, $\gamma^2 \sim 0.5$, we should have a modulation amplitude W_0 of the order of $0.02 \, E_F \sim 0.06$ meV, which is well within current technology.

4   Conclusions

We have shown that the transport in a semiconductor quantum wire is described as a Tomonaga-Luttinger liquid and for the lowest occupied subband, the interaction parameter K_w is about 0.7. By lowering the temperature, the conductance initially decreases by the interaction effect, then saturates at a length-dependent characteristic temperature. In a modulated quantum wire, we predicted that the backscattering current, I_{bs}, and the extra resistance ΔR exhibits power-law behavior $\Delta R \propto T^\alpha$ and $I_{bs} \propto \Omega^{\alpha+1}$ at high energies $k_B T \gg \frac{\hbar v_w}{L}$ or $\Omega \gg \frac{v_w}{L}$. Here an interaction-dependent exponent is $\alpha = K_w - 1$ for double barrier potential and $\alpha = K_w - 2$ for periodic potential near full-filling condition. At low energies, $k_B T \ll \frac{\hbar v_w}{L}$ and $\Omega \ll \frac{v_w}{L}$, the exponent of $\alpha = 0$ is independent of the interaction strength, which can be understood because the system's response is mainly determined by noninteracting contacts.

We are grateful to G. E. W. Bauer, Yu. V. Nazarov, T. Honda, T. Saku, V. V. Ponomarenko, and N. Nagaosa for stimulating discussions. Y. T. acknowledges financial support of the Dutch Foundation for Fundamental Research on Matter (FOM) and a hospitality by G. E. W. Bauer at Delft University of Technology. A. O. acknowledges financial support of the Royal Dutch Academy of Sciences (KNAW).

References

1. T. Honda, S. Tarucha, T. Saku, and Y. Tokura, Jpn J. Appl. Phys. **34**, L72 (1995).
2. S. Tarucha, T. Honda, and T. Saku, Solid State Commun. **98**, 413 (1995).
3. A. Yacoby, H. L. Stormer, N. S. Wingreen, L. N. Pfeiffer, K. W. Baldwin, and K. W. West, Phys. Rev. Lett. **77**, 4612 (1996).
4. C. Dekker, Phyics Today, May 22 (1999) and references therein.

5. M. Bockrath, D. H. Cobden, J. Lu, A. G. Rinzler, R. E. Smalley, L. Balents, and P. L. McEuen, Nature **397**, 598 (1999).

6. F. D. M. Haldane, J. Phys. C: Solid State Phys. **14**, 2585 (1981).

7. C. L. Kane and M. P. A. Fisher, Phys. Rev. Lett. **68**, 1220 (1992).

8. D. Kaufman, Y. Berk, B. Dwir, A. Rudra, A. Palevski, and E. Kapon, Phys. Rev. B **59**, R10433 (1999).

9. A. Y. Alekseev and V. V. Cheianov, Phys. Rev. B **57**, R6834 (1998).

10. M. Rother, W. Wegscheider, M. Bichler, and G. Abstreiter, 24th Int. Conf. on the physics of Semiconductors, Ed. by D. Gershoni, World Scientific (1999).

11. K. J. Thomas, J. T. Nicholls, N. J. Appleyard, M. Pepper, M. Y. Simmons, D. R. Mace, W. R. Tribe, and D. A. Ritchie, Phys. Rev. B **58**, 4846 (1998).

12. D. L. Maslov and M. Stone, Phys. Rev. B **52**, R5539 (1995).

13. V. V. Ponomarenko, Phys. Rev. B **52**, R8666 (1995).

14. I. Safi and H. J. Shultz, Phys. Rev. B **52**, R17040 (1995).

15. M. Ogata and H. Fukuyama, Phys. Rev. Lett. **73**, 468 (1994).

16. D. L. Maslov, Phys. Rev. B **52**, R14368 (1995).

17. S. Tarucha, *et al*, to be submitted.

18. K. A. Matveev and L. I. Glazman, Phys. Rev. Lett. **70**, 990 (1993).

19. A. Kawabata and T. Brandes, J. Phys. Soc. Jpn **65**, 3712 (1996).

20. N. P. Sandler and D. L. Maslov, Phys. Rev. B **55**, 13808 (1997).

21. A. Kawabata, J. Phys. Soc. Jpn **65**, 30 (1996).

22. A. Shimizu, J. Phys. Soc. Jpn **65**, 1162 (1996).

23. I. Safi, Phys. Rev. B **55**, R7331 (1997).

24. A. Furusaki and N. Nagaosa, Phys. Rev. B **54**, R5239 (1996).

25. C. L. Kane and M. P. A. Fisher, Phys. Rev. B **46**, 15233 (1992).

26. A. Furusaki and N. Nagaosa, Phys. Rev. B **47**, 3827 (1993).

27. I. Safi, Phys. Rev. B **56**, R12691 (1997).

28. A. A. Odintsov, Y. Tokura, and S. Tarucha, Phys. Rev. B **56**, R12729 (1997).

29. S. Eggert, cond-mat/9909001.

30. L. I. Glazman, F. W. J. Hekking, and A. I. Larkin, cond-mat/9903181.

31. Y. V. Nazarov, A. A. Odintsov, and D. V. Averin, Europhys. Lett. **37**, 213 (1997).

32. Y. Tokura, A. A. Odintsov, and S. Tarucha, 24th Int. Conf. on the physics of Semiconductors, Ed. by D. Gershoni, World Scientific (1999). Y. Tokura, A. A. Odintsov, and S. Tarucha, to be published.

33. V. V. Ponomarenko and N. Nagaosa, Phys. Rev. Lett. **79**, 1714 (1997).

Correlated Electrons in Carbon Nanotubes

Arkadi A. Odintsov[1] and Hideo Yoshioka[2]

[1] Department of Applied Physics and DIMES, Delft University of Technology, 2628 CJ Delft, The Netherlands
and Nuclear Physics Institute, Moscow State University, Moscow 119899, Russia
[2] Department of Physics, Nagoya University, Nagoya 464-8602, Japan

Abstract. Single-wall carbon nanotubes are almost ideal systems for the investigation of exotic many-body effects due to non-Fermi liquid behavior of interacting electrons in one dimension. Recent theoretical and experimental results are reviewed with a focus on electron correlations. Starting from a microscopic lattice model we derive an effective phase Hamiltonian for conducting single-wall nanotubes with arbitrary chirality. The parameters of the Hamiltonian show very weak dependence on the chiral angle, which makes the low-energy physics of conducting nanotubes universal. The temperature-dependent resistivity and frequency-dependent optical conductivity of nanotubes with impurities are evaluated within the Luttinger-like model. Localization effects are studied. In particular, we found that intra-valley and inter-valley electron scattering can not coexist at low energies. Low-energy properties of clean nanotubes are studied beyond the Luttinger liquid approximation. The strongest Mott-like electron instability occurs at half filling. In the Mott insulating phase electrons at different atomic sublattices form characteristic bound states. The energy gaps occur in all modes of elementary excitations and estimate at $0.01 - 0.1$ eV. We finally discuss observability of the Mott insulating phase in transport experiments. The accent is made on the charge transfer from external electrodes which results in a deviation of the electron density from half-filling.

1 Introduction

Single-wall carbon nanotubes (SWNTs) are cylindrical fullerene structures with diameters in the nanometer range and lengths of few micrometers [1]. Experimental demonstration of electron transport through SWNTs [2,3] has been followed by observations of atomic structure [4–6], one-dimensional van Hove singularities [4,5], standing electron waves [7] and electron correlations [8–10] in these systems. Moreover, first prototypes of SWNT functional devices - diodes [10,11] and field effect transistors [12] - have been demonstrated recently.

On one hand, SWNTs can be viewed as giant macromolecules whose properties can be learned from the first principle calculations. On the other hand, SWNTs are perfect one-dimensional (1D) model systems to be studied by methods of the solid state theory. This somewhat reductionist but insightful approach provides a reasonable description for the bulk of experimental data obtained up to date.

On a single-particle level physical properties of SWNTs are determined by their geometry. Depending on the wrapping vector, SWNTs can either be 1D metals or semiconductors with the energy gap in sub-electronvolt range. This has

98   A. Odintsov and H. Yoshioka

been confirmed by direct observation of 1D van Hove singularities in scanning tunneling microscopy experiments [4,5].

In this paper we address the role of the Coulomb interaction in 1D SWNTs and review recent results on electron correlation effects. Away from half-filling electron correlations are well described by Luttinger-like models [13–15]. In particular, the non-Fermi liquid ground state of the system is characterized by a power-law suppression of the density of electronic states near the Fermi level. This effect has been observed in single- [8] and, presumably, multi-wall [9] nanotubes, as well as in junctions between metallic SWNTs [10]. Transport properties of metallic nanotubes with impurities are affected by the Coulomb interaction as well [16].

The low-energy properties of SWNTs are different at half-filling due to the umklapp scattering. The latter is coupled to the strongly interacting total charge mode. This makes the umklapp scattering a strongly relevant perturbation. As a result, the Mott-like electron instability occurs in the electronic spectrum of SWNTs and energy gaps open in all modes of the elementary excitations [13,15]. The transition to the Mott insulating phase should manifest itself by an increased resistivity of half-filled SWNTs at low temperatures.

The plan of the paper is as follows. In Section 2 we derive an effective low-energy model for metallic SWNTs with arbitrary chirality, which naturally includes one-electron backscattering on impurities as well as two-electron umklapp scattering at half-filling. We discuss the non-Fermi liquid properties of clean SWNTs within the Luttinger-like model in Section 3 and evaluate electron transport in SWNTs with impurities in Section 4. The effect of interactions beyond the Luttinger model is analyzed within the renormalization group scheme in Section 5. The Mott-like electron instability is further investigated in Section 6 using the self-consistent harmonic approximation. Finally, we discuss observability of the Mott-insulating phase in experiments (Section 7).

2   Universal Model of Metallic Nanotubes

2.1   Microscopic Theory

Structurally uniform SWNTs can be characterized by the wrapping vector $w = N_+ a_+ + N_- a_-$ given by the linear combination of primitive lattice vectors $a_\pm = (\pm 1, \sqrt{3})a/2$, with $a \approx 0.246$ nm (Fig. 1). It is natural to separate non-chiral armchair ($N_+ = N_-$) and zig-zag ($N_+ = -N_-$) nanotubes from their chiral counterparts. Recent scanning tunneling microscopy studies [4,5] have revealed that individual SWNTs are generally chiral. According to the single-particle model, the nanotubes with $N_+ - N_- = 0$ mod 3 have gapless energy spectrum and are therefore conducting; otherwise, the energy spectrum is gapped and SWNTs are insulating.

We consider metallic SWNT whose axis x' forms an angle $\chi = \arctan[(N_- - N_+)/\sqrt{3}(N_+ + N_-)]$ with the direction of the chains of carbon atoms (x axis in Fig. 1). We expand the standard single-particle Hamiltonian [18] H_k for electrons

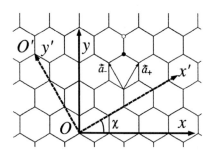

Fig. 1. Graphite lattice consists of two atomic sublattices $p = +, -$ denoted by filled and open circles. SWNT at the angle χ to x axis can be formed by wrapping the graphite sheet along $\boldsymbol{w} = OO'$ vector.

on two atomic sublattices $p = \pm$ of a graphite sheet (Fig. 1) near the Fermi points αK (with the valley index $\alpha = \pm$, and $\boldsymbol{K} = (4\pi/3a, 0)$) to the lowest order in $\boldsymbol{q} = \boldsymbol{k} - \alpha \boldsymbol{K} = q(\cos\chi, \sin\chi)$. Introducing slowly varying Fermi fields $\psi_{p\alpha s}(x') = L^{-1/2} \sum_{q=2\pi n/L} e^{iqx'} a_{p\alpha s}(\boldsymbol{q} + \alpha\boldsymbol{K})$, we obtain,

$$H_k = -iv \sum_{p\alpha s} \alpha e^{-ip\alpha\chi} \int dx' \psi^\dagger_{p\alpha s} \partial_{x'} \psi_{-p\alpha s}, \tag{1}$$

where $v \approx 8.1 \times 10^5$ m/s is the Fermi velocity, and $s = \pm$ is the electron spin.

The kinetic term can be diagonalized by the unitary transformation

$$\psi_{p\alpha s} = \frac{1}{\sqrt{2}} e^{-ip\alpha\chi/2} \sum_{r=\pm} (r\alpha)^{\frac{1-p}{2}} \varphi_{r\alpha s} \tag{2}$$

to the basis $\varphi_{r\alpha s}$ of right- $(r = +)$ and left- $(r = -)$ and moving electrons.

The Coulomb interaction has the form

$$H_{int} = \frac{1}{2} \sum_{pp', \{\alpha_i\}, ss'} V_{pp'}(2\bar{\alpha}K) \int dx' \psi^\dagger_{p\alpha_1 s} \psi^\dagger_{p'\alpha_2 s'} \psi_{p'\alpha_3 s'} \psi_{p\alpha_4 s}, \tag{3}$$

with the matrix elements $V_{pp'}(2\bar{\alpha}K)$ corresponding to the amplitudes of intra- $(p = p')$ and inter- $(p = -p')$ sublattice intra- $(\bar{\alpha} = 0)$ and inter- $(\bar{\alpha} = \pm 1)$ valley scattering (here $\bar{\alpha} = (\alpha_1 - \alpha_4)/2 = (\alpha_3 - \alpha_2)/2$). In Eq. (3) we assume that the fields ψ are varying slowly on the scale of the screening radius R_s of the Coulomb interaction determined by the distance to a gate electrode. This corresponds to a contact-interaction approximation. Equation (3) is therefore valid at large length scale $x' \gg R_s$ and low electron energy $E \ll \hbar v/R_s$.

The dominant contribution to the intra-valley scattering amplitudes $V_{pp'}(0)$ comes from the long range component of the Coulomb interaction, $V_{pp}(0) \simeq V_{p-p}(0) \simeq e^2/C \simeq (2e^2/\kappa) \ln(R_s/R)$, C being the capacitance of SWNT per unit length [13]. The differential part $\Delta V(0) = V_{pp}(0) - V_{p-p}(0)$ of intra-valley

Table 1. Scattering amplitudes $\Delta V(0)$, $V_{pp}(2K)$, $V_{p-p}(2K)$ in units $ae^2/2\pi\kappa R$ for all SWNTs with $2R/a = 4 - 7$. The cutoff parameter $a_0 = 0.526a$ is obtained from the requirement that the on-site interaction in the original tight-binding model is equal to the difference between the ionization potential and electron affinity of sp^2 hybridized carbon [19].

a_0/a	$\Delta V(0)$	$V_{pp}(2K)$	$|V_{p-p}(2K)|$
0.4	$0.44265 - 0.44274$	$0.97060 - 0.97095$	$0.6 - 2.2 \times 10^{-3}$
0.526	$0.17378 - 0.17395$	$0.53549 - 0.53561$	$0.5 - 1.6 \times 10^{-3}$
0.7	$0.04880 - 0.04895$	$0.24778 - 0.24797$	$0.3 - 1.5 \times 10^{-3}$

scattering as well as the intra-sublattice inter-valley $V_{pp}(2K)$ are estimated at $\Delta V(0), V_{pp}(2K) \sim ae^2/\kappa R$. Despite $\Delta V(0), V_{pp}(2K)$ being much smaller than $V_{pp}(0)$, they cause non-Luttinger terms in the low-energy Hamiltonian which will be important in the further analysis. The matrix elements $V_{pp'}(2\bar{a}K)$ have been evaluated numerically for all SWNTs with radii R in the range $2R/a = 4 - 7$ ($2R/a = 5.5$ for (10,10) SWNTs). We found that dimensionless amplitudes $2\pi\kappa R[\Delta V(0), V_{pp}(2K)]/ae^2$ show very weak dependence on the radius of SWNT and its chiral angle (see Table 1). The results are sensitive to the value the short distance cutoff $a_0 \sim a$ of the Coulomb interaction.

The inter-sublattice inter-valley scattering amplitude $V_{p-p}(2K)$ is almost three orders of magnitude smaller than $\Delta V(0), V_{pp}(2K)$. This is due to the C_3 symmetry of a graphite lattice ($V_{p-p}(2K) = 0$ for a plane graphite sheet). The matrix elements $V_{p-p}(2K)$ are generally complex due to the asymmetry of effective 1D inter-sublattice interaction potential (the matrix elements are real for symmetric zig-zag and armchair SWNTs). Let us note that after the unitary transformation (2) of the Hamiltonian $H = H_k + H_{int}$, the chiral angle χ enters *only* into the inter-sublattice inter-valley scattering matrix elements $V_{p-p}(2K)$. Due to the smallness of these matrix elements, the low-energy properties of chiral SWNTs are expected to be virtually independent of the chiral angle.

2.2   Bosonization

The transformed Hamiltonian H can be bosonized by introducing the phase representation of the Fermi fields [14,15],

$$\varphi_{r\alpha s} = \frac{\eta_{r\alpha s}}{\sqrt{2\pi\tilde{a}}} \exp\left[ir q_F x' + \frac{ir}{2}\{\theta_{\alpha s} + r\phi_{\alpha s}\}\right]. \tag{4}$$

The phase variables $\theta_{\alpha s}, \phi_{\alpha s}$ are further decomposed into symmetric $\delta = +$ and antisymmetric $\delta = -$ modes of the charge ρ and spin σ excitations, $O_{\alpha s} = O_{\rho+} + sO_{\sigma+} + \alpha O_{\rho-} + \alpha s O_{\sigma-}$, $O = \theta, \phi$. The bosonic fields satisfy the commutation relation, $[\theta_{j\delta}(x_1), \phi_{j'\delta'}(x_2)] = i(\pi/2)\text{sign}(x_1-x_2)\delta_{jj'}\delta_{\delta\delta'}$. The Majorana fermions $\eta_{r\alpha s}$ are introduced to ensure correct anticommutation rules for different species r, α, s of electrons, and satisfy $[\eta_{r\alpha s}, \eta_{r'\alpha's'}]_+ = 2\delta_{rr'}\delta_{\alpha\alpha'}\delta_{ss'}$. The quantity $q_F =$

$\pi n/4$ is related to the deviation n of the average electron density from half-filling, and $\tilde{a} \sim a$ is the parameter of the exponential ultraviolet cutoff.

Neglecting the inter-sublattice inter-valley scattering we arrive at the universal phase Hamiltonian of metallic SWNTs,

$$H = \sum_{j=\rho,\sigma} \sum_{\delta=\pm} \frac{v_{j\delta}}{2\pi} \int dx' \left\{ K_{j\delta}^{-1}(\partial_{x'}\theta_{j\delta})^2 + K_{j\delta}(\partial_{x'}\phi_{j\delta})^2 \right\} + \frac{1}{2(\pi\tilde{a})^2} \int dx'$$

$$\{[\Delta V(0) - V_{pp}(2K)][\cos(4q_F x' + 2\theta_{\rho+})\cos 2\theta_{\sigma+} - \cos 2\theta_{\rho-}\cos 2\theta_{\sigma-}]$$
$$-\Delta V(0)\cos(4q_F x' + 2\theta_{\rho+})\cos 2\theta_{\rho-} + \Delta V(0)\cos(4q_F x' + 2\theta_{\rho+})\cos 2\theta_{\sigma-}$$
$$-\Delta V(0)\cos 2\theta_{\sigma+}\cos 2\theta_{\rho-} + \Delta V(0)\cos 2\theta_{\sigma+}\cos 2\theta_{\sigma-}$$
$$-V_{pp}(2K)\cos(4q_F x' + 2\theta_{\rho+})\cos 2\phi_{\sigma-} + V_{pp}(2K)\cos 2\theta_{\sigma+}\cos 2\phi_{\sigma-}$$
$$+V_{pp}(2K)\cos 2\theta_{\rho-}\cos 2\phi_{\sigma-} + V_{pp}(2K)\cos 2\theta_{\sigma-}\cos 2\phi_{\sigma-}\}, \tag{5}$$

$v_{j\delta} = v\sqrt{A_{j\delta}B_{j\delta}}$ and $K_{j\delta} = \sqrt{B_{j\delta}/A_{j\delta}}$ being the velocities and interaction parameters for different modes j,δ of excitations. The parameters $A_{j\delta}$, $B_{j\delta}$ are given by $A_{\rho+} = 1 + [8\bar{V}(0) - \Delta V(0)/2 - V_{pp}(2K)]/2\pi v$, $A_{\nu\delta} = 1 - [\Delta V(0)/2 + \delta V_{pp}(2K)]/2\pi v$, $B_{\nu\delta} = 1 + \Delta V(0)/4\pi v$, with $\bar{V}(0) = [V_{pp}(0) + V_{p-p}(0)]/2$. The renormalization of the parameters $K_{j\delta}$, $v_{j\delta}$ by the Coulomb interaction is strongest in $\rho+$ mode. Assuming $\kappa = 1.4$ [14] $R = 0.7$ nm, and $R_s = 100$ nm we obtain $K_{\rho+} \simeq 0.2$. The interaction in the other modes is weak: $K_{j\delta} = 1+O(a/R)$.

Let us note that the Hamiltonian (5) has the same form as the phase Hamiltonian for a two-leg Hubbard-type ladder [20], provided that the difference in definitions of the fields θ_{j-} and ϕ_{j-} ($j = \rho, \sigma$) in terms of densities of right- and left-movers in two energy bands is taken into account.

2.3 Impurity Scattering

Disorder in the atomic potential is described by the Hamiltonian [21,16], $H_{imp} = \sum_{p\alpha\alpha's} \int dx V_{p\bar{\alpha}}(x)\psi_{pas}^{\dagger}\psi_{p\alpha's}$. Here the impurity potential $V_p(x)$ at the sublattice $p = \pm$ is decomposed into intra-valley ($\bar{\alpha} \equiv (\alpha' - \alpha)/2 = 0$) and inter-valley ($\bar{\alpha} = \pm 1$) scattering components. First, we transform the Hamiltonian H_{imp} to the basis of right- and left-movers ($r = \pm$), see Eq. (2). When the range of the impurity potential is much larger than the lattice constant, the backward scattering ($r \to -r$) is ineffective [21]. We consider the case of short range impurity potential and retain backscattering terms in the Hamiltonian. The forward scattering is discarded because it does not contribute to the transport.

In the limit of weak impurity potential, the interaction between the electrons and the impurities can be parameterized by uncorrelated Gaussian random fields, $\eta(x)$ and $\xi(x)$ expressing the intra-valley and the inter-valley backward scattering, respectively. The fields satisfy $\langle\eta(x)\eta(x')\rangle_{imp} = D_1\delta(x - x')$ and $\langle\xi(x)\xi^*(x')\rangle_{imp} = D_2\delta(x - x')$, where $\langle\cdots\rangle_{imp}$ is the configurational average. The factors D_1 and D_2 are given by v/τ_1 and v/τ_2 with the scattering time τ_1 (τ_2) due to the intra-valley (inter-valley) backscattering. The Hamiltonian of impurities is given by

$$H_{imp} = \int dx \eta(x) \sum_{r\alpha s} \psi^\dagger_{r\alpha s} \psi_{-r\alpha s} + \int dx \left\{ \xi(x) \sum_{rs} \psi^\dagger_{r+s} \psi_{-r-s} + h.c. \right\}. \quad (6)$$

Note that the intra-valley (inter-valley) backward scattering is parameterized by a real $\eta(x)$ (complex $\xi(x)$) field. The Hamiltonian $H_{imp} = H^1_{imp} + H^2_{imp}$ is expressed in terms of the phase variables as follows,

$$H^1_{imp} = \frac{i\sigma_z}{2\pi\tilde{a}} \int dx \eta(x) \sum_{r\alpha s} r\alpha \exp\left(-2irq_F x\right)$$

$$\times \exp\left\{-ir(\theta_{\rho+} + s\theta_{\sigma+} + \alpha\theta_{\rho-} + \alpha s\theta_{\sigma-})\right\}, \quad (7)$$

$$H^2_{imp} = \frac{-i\sigma_y}{2\pi\tilde{a}} \int dx \sum_{rs} r \exp\left\{-ir(2q_F x + \theta_{\rho+} + s\theta_{\sigma+})\right\}$$

$$\times \left[\xi(x) \exp\left\{-i(\phi_{\rho-} + s\phi_{\sigma-})\right\} + h.c.\right]. \quad (8)$$

with the Pauli matrices σ_y, σ_z originating from the Majorana representation of the Fermi fields (4).

3   Luttinger Model Limit

The phase Hamiltonian (5) of clean metallic SWNT consists of a part quadratic in bosonic fields describing scattering of electrons within the same branch of the spectrum and the non-quadratic part describing inter-branch scattering. In the Luttinger model limit one neglects the differential part $\Delta V(0)$ of intra-valley scattering and the inter-valley scattering $V_{pp'}(2K)$. In this case the Hamiltonian (5) reads,

$$H_L = \sum_{j=\rho,\sigma} \sum_{\delta=\pm} \frac{v_{j\delta}}{2\pi} \int dx' \left\{ K^{-1}_{j\delta}(\partial_{x'}\theta_{j\delta})^2 + K_{j\delta}(\partial_{x'}\phi_{j\delta})^2 \right\}, \quad (9)$$

with parameters $v_{j\delta} = v\sqrt{A_{j\delta}}$ and $K_{j\delta} = 1/\sqrt{A_{j\delta}}$ determined by $A_{\rho+} = 1 + 4\bar{V}(0)/\pi v$, $A_{j\delta} = 1$ for $(j\delta) \neq (\rho+)$. The parts of the Hamiltonian (9) in the four sectors of excitations $(j\delta)$ are decoupled. In particular, using the relations $\rho = (2/\pi)\partial_x\theta_{\rho+}$, $j = (2v/\pi)\partial_x\phi_{\rho+}$ for the charge density $e\rho$ and the electric current ej ($e > 0$) in SWNT, the Hamiltonian (9) in the total charge $\rho+$ sector can be rewritten as follows,

$$H_{\rho+} = \frac{1}{\nu} \int dx' \left[\frac{1}{2}\rho^2 + \frac{1}{2}\left(\frac{j}{v}\right)^2 \right] + \frac{1}{2} \int dx'dx'' \rho(x')V(x' - x'')\rho(x''), \quad (10)$$

(in Eq. (9) the contact interaction approximation $V(x) = \bar{V}(0)\delta(x)$ has been made). In Eq. (10) $\nu = 4/\pi v$ is the density of electronic states in SWNT; the first term describes the kinetic energy of right and left moving electrons and the second corresponds to the Hartree part of the interaction.

Electronic properties of SWNTs in the Luttinger limit have been investigated in e.g. Refs. [13,14]. The density of electronic states near the Fermi level is suppressed in a power-law fashion. This is a signature of the orthogonality catastrophe which occurs due to the non-fermionic (bosonic) nature of low-energy excitations in SWNTs. As a result, the differential conductance in the tunneling regime shows a power-law dependence on temperature, $dI/dV \propto T^\alpha$, for $eV \ll T$ and voltage, $dI/dV \propto V^\alpha$, for $eV \gg T$. The exponent α is given by $\alpha = (K_{\rho+}^{-1} + K_{\rho+} - 2)/8$ for the tunneling from a metallic electrode into the "bulk" of SWNT, $\alpha = (K_{\rho+}^{-1} - 1)/4$ for the tunneling into the end of SWNT, and $\alpha = (K_{\rho+}^{-1} - 1)/2$ for the end-to-end tunneling in nanotube heterojunctions [10]. Moreover, scaled differential conductance, $(dI/dV)/T^\alpha$, is a universal function of the ratio eV/T. The collapse of the data taken at different temperatures to a single universal curve provides a comprehensive check of validity of the Luttinger model for SWNTs [8,10].

4   Effect of Impurities

In this section, the transport properties of metallic SWNTs with impurities are studied in the Luttinger model limit [16]. The dynamical conductivity $\sigma(\omega)$ is expressed via the memory function $M(\omega)$ as follows [22],

$$\sigma(\omega) = \frac{-i\chi(0)}{\omega + M(\omega)}, \quad M(\omega) = \frac{(\langle F; F \rangle_\omega - \langle F; F \rangle_{\omega=0})/\omega}{-\chi(0)}, \qquad (11)$$

where $\langle A; A \rangle_\omega \equiv -i \int dx \int_0^\infty dt e^{(i\omega - \eta)t} \langle [A(x,t), A(0,0)] \rangle$ with $\eta \to +0$, $\langle \cdots \rangle$ denotes the thermal average with respect to the Hamiltonian $H = H_L + H_{imp}$, $F = [j, H]$ with j being the current operator, and $\chi(0) = \langle j; j \rangle_{\omega=0} = -4v/\pi$.

The temperature dependence of the resistivity is given by,

$$\rho = \rho_{B0} \frac{\Gamma^2((K_{\rho+} + 3)/4)}{\Gamma((K_{\rho+} + 3)/2)} \left(\frac{2\pi T}{\omega_F}\right)^{(K_{\rho+} - 1)/2}, \qquad (12)$$

where $\rho_{B0} = (\pi/2) \sum_{i=1,2} (v\tau_i)^{-1}$, is the resistivity of the non-interacting system in the Born approximation, $\omega_F = v/\tilde{a}$, and $\Gamma(z)$ is the gamma function. It is remarkable that ρ/ρ_{B0} is independent of the scattering strength. The repulsive Coulomb interaction ($K_{\rho+} < 1$) leads to an enhancement of the resistivity at low temperatures. For typical nanotubes with $K_{\rho+} \simeq 0.2$, the resistivity scales as $\rho \propto T^{-0.4}$.

The optical conductivity $\mathrm{Re}\sigma(\omega)$ behaves like $\omega^{(K_{\rho+}-5)/2}$ at high frequencies $\omega \gg T$ and as $\omega^{(1-K_{\rho+})/2}$ at low frequencies. The position of the peak in $\mathrm{Re}\sigma(\omega)$ is given by,

$$\frac{\omega}{\omega_F} \sim \left[\frac{2}{\omega_F}(\frac{1}{\tau_1} + \frac{1}{\tau_2})\frac{\tan\{\pi(1 - K_{\rho+})/4\}}{\Gamma((K_{\rho+} + 3)/2)}\right]^{2/(3-K_{\rho+})}, \qquad (13)$$

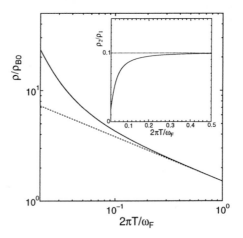

Fig. 2. Temperature dependence of the resistivity ρ of SWNT with the following parameters, $\pi \omega_F \tau_1 = 300$ and $\pi \omega_F \tau_2 = 600$. The solid (dotted) line corresponds to RG analysis (perturbation theory). Inset : The ratio ρ_2/ρ_1 as a function of temperature.

We will further study localization of electrons which corresponds to the pinning of the phase fields $\theta_{j\delta}$, $\phi_{j\delta}$. The localization is described within the renormalization group (RG) formalism. RG equations can be derived by assuming scale invariance of correlation functions along the lines of Ref. [23],

$$(\mathcal{D}_1)' = \{3 - (K_{\rho+} + K_{\sigma+} + K_{\rho-} + K_{\sigma-})/2\}\,\mathcal{D}_1, \tag{14}$$

$$(\mathcal{D}_2)' = \{3 - (K_{\rho+} + K_{\sigma+} + K_{\rho-}^{-1} + K_{\sigma-}^{-1})/2\}\,\mathcal{D}_2, \tag{15}$$

$$(K_{j+})' = - (\mathcal{D}_1/X_1 + \mathcal{D}_2/X_2)\,K_{j+}^2 u_{j+}, \tag{16}$$

$$(u_{j+})' = - (\mathcal{D}_1/X_1 + \mathcal{D}_2/X_2)\,K_{j+} u_{j+}^2, \tag{17}$$

$$(K_{j-})' = - (\mathcal{D}_1 K_{j-}^2/X_1 - \mathcal{D}_2/X_2)\,u_{j-}, \tag{18}$$

$$(u_{j-})' = - (\mathcal{D}_1 K_{j-}/X_1 + \mathcal{D}_2 K_{j-}^{-1}/X_2)\,u_{j-}^2, \tag{19}$$

where ( )$'$ denotes $d/d\ell$ with $d\ell = d\ln(\tilde{a}/a)$ ($\tilde{a}$ is the new lattice constant), $X_1 = u_{\rho+}^{K_{\rho+}/2} u_{\sigma+}^{K_{\sigma+}/2} u_{\rho-}^{K_{\rho-}/2} u_{\sigma-}^{K_{\sigma-}/2}$, $X_2 = u_{\rho+}^{K_{\rho+}/2} u_{\sigma+}^{K_{\sigma+}/2} u_{\rho-}^{1/2K_{\rho-}} u_{\sigma-}^{1/2K_{\sigma-}}$ and $j = \rho$ or σ. The initial conditions for the above RG equations are as follows, $\mathcal{D}_i(0) = D_i \tilde{a}/(\pi v^2)$, $K_{\rho+}(0) = u_{\rho+}^{-1} = K_{\rho+}$, and $K_{\sigma+}(0) = K_{\rho-}(0) = K_{\sigma-}(0) = u_{\sigma+}(0) = u_{\rho-}(0) = u_{\sigma-}(0) = 1$.

The temperature dependence of the resistivity, Fig. 2, can be evaluated by solving the RG equations. The enhancement of the resistivity at low-temperatures is due to the electron localization. Solution of RG equations shows that the intra-valley (inter-valley) scattering pins the phases, $\theta_{\rho+}, \theta_{\sigma+}, \theta_{\rho-}$, and $\theta_{\sigma-}$ ($\theta_{\rho+}, \theta_{\sigma+}, \phi_{\rho-}$, and $\phi_{\sigma-}$). Since the conjugate variables, $\theta_{\rho-}$ and $\phi_{\rho-}$, or $\theta_{\sigma-}$ and $\phi_{\sigma-}$ cannot be pinned at the same time, the localization due to two kinds of the scattering cannot occur simultaneously. For the parameters of Fig. 2, the intra-valley scattering dominates over the inter-valley scattering at low temperatures (see inset of Fig. 2).

The high frequency behavior of the optical conductivity is not modified by the effects of the localization. The power law behavior at low frequencies survives due to a finite density of states at the Fermi energy (similarly to the non-interacting case [24]).

5   Effect of Interactions Beyond the Luttinger Model

Until now the Coulomb interaction in SWNTs has been treated within the Luttinger model. In order to describe the interaction effects beyond this approximation, one has to consider the full Hamiltonian (5). The low energy properties of Eq. (5) can be investigated by the RG method (see Ref. [23] for details). At half-filling, $q_F = 0$, we obtain the following RG equations,

$$(K_{\rho+})' = -(K_{\rho+}^2/8)(y_1^2 + y_2^2 + y_3^2 + y_7^2) \ , \tag{20}$$

$$(K_{\sigma+})' = -(K_{\sigma+}^2/8)(y_1^2 + y_5^2 + y_6^2 + y_8^2) \ , \tag{21}$$

$$(K_{\rho-})' = -(K_{\rho-}^2/8)(y_2^2 + y_4^2 + y_6^2 + y_9^2) \ , \tag{22}$$

$$(K_{\sigma-})' = -(K_{\sigma-}^2/8)(y_3^2 + y_4^2 + y_5^2) + (1/8)(y_7^2 + y_8^2 + y_9^2) \ , \tag{23}$$

$$(y_1)' = \{2 - (K_{\rho+} + K_{\sigma+})\} y_1 - (y_2 y_6 + y_3 y_5 + y_7 y_8)/4 \ , \tag{24}$$

$$(y_2)' = \{2 - (K_{\rho+} + K_{\rho-})\} y_2 - (y_1 y_6 + y_3 y_4 + y_7 y_9)/4 \ , \tag{25}$$

$$(y_3)' = \{2 - (K_{\rho+} + K_{\sigma-})\} y_3 - (y_1 y_5 + y_2 y_4)/4 \ , \tag{26}$$

$$(y_4)' = \{2 - (K_{\rho-} + K_{\sigma-})\} y_4 - (y_2 y_3 + y_5 y_6)/4 \ , \tag{27}$$

$$(y_5)' = \{2 - (K_{\sigma+} + K_{\sigma-})\} y_5 - (y_1 y_3 + y_4 y_6)/4 \ , \tag{28}$$

$$(y_6)' = \{2 - (K_{\sigma+} + K_{\rho-})\} y_6 - (y_1 y_2 + y_4 y_5 + y_8 y_9)/4 \ , \tag{29}$$

$$(y_7)' = \{2 - (K_{\rho+} + 1/K_{\sigma-})\} y_7 - (y_1 y_8 + y_2 y_9)/4 \ , \tag{30}$$

$$(y_8)' = \{2 - (K_{\sigma+} + 1/K_{\sigma-})\} y_8 - (y_1 y_7 + y_6 y_9)/4 \ , \tag{31}$$

$$(y_9)' = \{2 - (K_{\rho-} + 1/K_{\sigma-})\} y_9 - (y_2 y_7 + y_6 y_8)/4 \ . \tag{32}$$

The initial conditions for Eqs. (20)-(32) are $K_{j\delta}(0) = K_{j\delta}$, $y_1 = [\Delta V(0) - V_{pp}(2K_0)]/(\pi v)$, $y_2 = -y_3 = -y_5 = y_6 = -\Delta V(0)/(\pi v)$, $y_4 = [V_{pp}(2K) - \Delta V(0)]/(\pi v)$, $y_7 = -y_8 = -V_{pp}(2K)/(\pi v)$, and $y_9 = V_{pp}(2K)/(\pi v)$. In deriving the RG equations, the non-linear term $\cos 2\theta_{\sigma-} \cos 2\phi_{\sigma-}$ is omitted because this operator stays exactly marginal in all orders and is thus decoupled from the problem [14]. The RG equations away from half-filling can be obtained from Eqs. (20)-(32) by putting y_1, y_2, y_3 and y_7 to zero. Hereafter we concentrate on the case $N = 10$, $\kappa = 1.4$, $R_s = 100 \ nm$ and $a_0/a = 0.526$ and estimate the initial values of the RG parameters using Table 1.

Away from half-filling, the quantities $K_{\sigma+}$, $K_{\rho-}$, and $K_{\sigma-}^{-1}$ renormalize to zero and the coefficient of $\cos 2\theta_{\sigma+} \cos 2\theta_{\rho-}$ ($\cos 2\theta_{\sigma+} \cos 2\phi_{\sigma-}$ and $\cos 2\theta_{\rho-} \cos 2\phi_{\sigma-}$) tends to $-\infty$ (∞). As a result, the phases $\theta_{\sigma+}, \theta_{\rho-}$ and $\phi_{\sigma-}$ are pinned at $(\theta_{\sigma+}, \theta_{\rho-}, \phi_{\sigma-}) = (0, 0, \pi/2)$ or $(\pi/2, \pi/2, 0)$ so that the modes $\sigma\pm$ and $\rho-$ are gapped. In this case, the asymptotic behavior of the correlation functions at $x \to \infty$ is determined by the correlations of the gapless $\rho+$ mode, $\langle e^{in\theta_{\rho+}(x)}e^{-in\theta_{\rho+}(0)}\rangle \sim x^{-n^2 K_{\rho+}/2}$ and $\langle e^{im\phi_{\rho+}(x)}e^{-im\phi_{\rho+}(0)}\rangle \sim x^{-m^2/2K_{\rho+}}$ ($n = 1$ and 2 correspond to $2q_F$ and $4q_F$ density waves and $m = 1$ to a superconducting state). Since $K_{\rho+} \approx 0.2$, the $2q_F$ density wave correlations seem to be dominant. However, we found that the correlation functions of any $2q_F$ density wave decay exponentially at large distances due to the gapped modes. We therefore are looking for the four-particle correlations. The $4q_F$ density waves dominate over superconductivity for $K_{\rho+} < 1/2$ [25,26]. Such density wave states are given by the product of the charge $\rho_+(x)\rho_-(x)$ or spin $S_+(x)S_-(x)$ densities at different sublattices. Substituting the values of the pinned phases we observe that $\rho_+(x)\rho_-(x)$ vanishes, and the dominant state is the $4q_F$ spin density wave with correlation function $\langle S_+(x)S_-(x)S_+(0)S_-(0)\rangle \sim \cos 4q_F x/x^{2K_{\rho+}}$.

At half-filling the solution of the RG equations (20)-(32) indicates that the phase variables $\theta_{\rho+}$, $\theta_{\sigma+}$, $\theta_{\rho-}$, and $\phi_{\sigma-}$ are pinned and the all kinds of excitation are gapped. In other words, the ground state of the half-filled AN is a Mott insulator with spin gap. The same conclusion has been drawn from the model with short range interactions [27,28]. The pinned phases are given by $(\theta_{\rho+},\theta_{\sigma+},\theta_{\rho-},\phi_{\sigma-}) = (0,0,0,0)$ or $(\pi/2,\pi/2,\pi/2,\pi/2)$ since the first, second and 6-9-th non-linear terms in Eq. (5) scale to $-\infty$. The averages $\langle\rho_+(x)\rho_-(x)\rangle$ and $\langle S_+(x)S_-(x)\rangle$ are both finite, which indicates formation of bound states of electrons at different atomic sublattices.

The states derived from the present analysis are characteristic for the long range Coulomb interaction. In fact, for the on-site plus nearest neighbor interaction the dominant states correspond to the density waves at half-filling and to the superconducting state or the density waves away from half-filling [28].

6   Mott-Insulating Phase

To estimate the gaps in different modes of excitations at half-filling, we employ the self-consistent harmonic approximation which follows from Feynman's variational principle [29]. We consider a trial harmonic Hamiltonian of the form:

$$H_0 = \sum_{j\delta} \frac{v_{j\delta}}{2\pi} \int dx' \{ K_{j\delta}^{-1}[(\partial_{x'}\theta_{j\delta})^2 + (1 - \delta_{j\sigma}\delta_{\delta-})q_{j\delta}^2\theta_{j\delta}^2]$$
$$+ K_{j\delta}[(\partial_{x'}\phi_{j\delta})^2 + \delta_{j\sigma}\delta_{\delta-}q_{j\delta}^2\phi_{j\delta}^2]\}, \qquad (33)$$

$q_{j\delta}$ being variational parameters. By minimizing the upper estimate for the free energy $F^* = F_0 + \langle H - H_0\rangle_0$ one obtains the following self-consistent equations,

$$q_{\rho+}^2 = \frac{2K_{\rho+}}{\pi\tilde{a}^2 v_{\rho+}}c_{\rho+}[V_{pp}(2K) - \Delta V(0)]c_{\sigma+} + \Delta V(0)c_{\rho-} + V_{pp}(2K)d_{\sigma-}, \quad (34)$$

$$q_{\rho-}^2 = \frac{2K_{\rho-}}{\pi\tilde{a}^2 v_{\rho-}}c_{\rho-}\Delta V(0)c_{\rho+} + \Delta V(0)c_{\sigma+} - V_{pp}(2K)d_{\sigma-}, \quad (35)$$

$$q_{\sigma+}^2 = \frac{2K_{\sigma+}}{\pi\tilde{a}^2 v_{\sigma+}}c_{\sigma+}[V_{pp}(2K) - \Delta V(0)]c_{\rho+} + \Delta V(0)c_{\rho-} - V_{pp}(2K)d_{\sigma-}, \quad (36)$$

$$q_{\sigma-}^2 = \frac{2}{\pi\tilde{a}^2 K_{\sigma-} v_{\sigma-}}d_{\sigma-} V_{pp}(2K)\{c_{\rho+} - c_{\sigma+} - c_{\rho-}\}, \quad (37)$$

where $c_{j\delta} = \langle\cos 2\theta_{j\delta}\rangle_0 = \cos 2\theta_{j\delta}^{(m)}(\gamma\tilde{a}q_{j\delta})^{K_{j\delta}}$, $d_{\sigma-} = \langle\cos 2\phi_{\sigma-}\rangle_0 = \cos 2\phi_{\sigma-}^{(m)}$ $(\gamma\tilde{a}q_{\sigma-})^{1/K_{\sigma-}}$, $\langle...\rangle_0$ denotes averaging with respect to the trial Hamiltonian (33), and $\gamma \simeq 0.890$. Note that $\langle\cos 2\theta_{\sigma-}\rangle_0 = 0$, so that only the terms of the Hamiltonian (5) which scale to the strong coupling (see end of Section 5) contribute to Eqs. (34)-(37).

In the limiting case of interest, $|\Delta V(0)|, |V_{pp}(2K)| \ll v$ and $K_{\rho+} \ll 1$, the solution of Eqs. (34)-(37) can be found in a closed form, giving rise to the following estimates for the gaps $\Delta_{j\delta} = v_{j\delta}q_{j\delta}$ in the energy spectra,

$$\Delta_{\rho+} = \frac{v_{\rho+}}{\gamma\tilde{a}}\left(\frac{2\gamma^2 V_{\rho+}}{\pi v_{\rho+}}\right)^{1/(1-K_{\rho+})} \quad (38)$$

$$\Delta_{\rho-} = \frac{|\Delta V(0)|}{V_{\rho+}}\Delta_{\rho+} \quad (39)$$

$$\Delta_{\sigma+} = \frac{|V_{pp}(2K_0) - \Delta V(0)|}{V_{\rho+}}\Delta_{\rho+} \quad (40)$$

$$\Delta_{\sigma-} = \frac{|V_{pp}(2K_0)|}{V_{\rho+}}\Delta_{\rho+} \quad (41)$$

with $V_{\rho+} = \{[\Delta V(0) - V_{pp}(2K)]^2 + [\Delta V(0)]^2 + [V_{pp}(2K)]^2\}^{1/2}$. In the above expressions we used the approximation, $v/v_{\rho+} = K_{\rho+}$ and $v/v_{j\delta} = K_{j\delta} = 1$ for $\sigma\pm$ and $\rho-$ modes. The formulae (38)-(41) indicate that the largest gap occurs in the $\rho+$ mode, albeit all four gaps are of the same order for realistic values of the matrix elements (see Table 1). The gaps decrease as $\Delta_{j\delta} \propto (1/R)^{1/(1-K_{\rho+})} \simeq 1/R^{5/4}$ with the nanotube radius.

In Figure 3 we present numerical results for the gaps $\Delta_{j\delta}$ for the short distance cutoff of the Coulomb interaction $a_0 = 0.526a$. The data for somewhat larger and somewhat smaller values of a_0 indicate possible variations of the gap $\Delta_{\rho+}$ due to the uncertainty in the cutoff. The gaps can be roughly estimated at $\Delta_{j\delta} \sim 0.01 - 0.1$ eV for typical SWNTs with $R \simeq 0.7$ nm.

The resistivity ρ of metallic SWNTs increases exponentially $\rho \propto \exp(\Delta_{\rho+}/T)$ at low temperatures $T \ll \Delta_{\rho+}$. On the other hand, at high temperatures, $T \gg \Delta_{\rho+}$, perturbation theory with respect to the non-linear terms of the Hamiltonian (5) gives a power-law behavior of the resistivity $\rho \sim T^{2K_{\rho+}-1}/N^2$ due to umklapp scattering at half-filling. Note that the power-law is different from that governing the impurity scattering, Eq. (12). A resonant increase of the resistivity of SWNTs at half-filling is a characteristic signature of the Mott insulating phase.

Due to the gaps in the spectrum of bosonic elementary excitations, the electronic density of states should disappear in the subgap region and display fea-

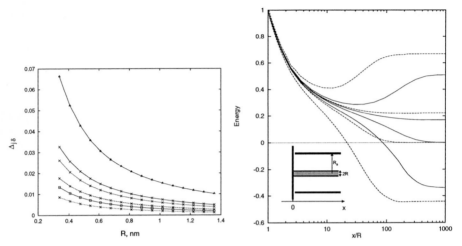

Fig. 3. The energy gaps $\Delta_{j\delta}$ for the modes $\rho+$, $\sigma-$, $\sigma+$, $\rho-$ at $a_0 = 0.526a$ (lines marked by crosses, from top to bottom) and for the mode $\rho+$ at $a_0 = 0.4a$ (triangles) and at $a_0 = 0.7a$ (squares). The energy is in units $\hbar v/\tilde{a} \simeq 2.16$ eV for $\tilde{a} = a$.

Fig. 4. The energy ΔE of the gapless point (half filling) with respect to the Fermi level (units of $\Delta W/(2e^2\nu/\kappa)$). The curves from top to bottom at $x/R = 1000$ correspond to $eV_g/\Delta W = -2, 0, 1, 3$. The screening radius $R_s/R = 75, 300$ for dashed and solid curves corresponds approximately to 50 nm and 200 nm for (10,10) SWNTs. The Coulomb interaction is characterized by $\nu \bar{V}(0)/\ln(R_s/R) = 5$. Inset: layout of the system.

tures at the gap frequencies and their harmonics. These signatures should be observable by means of the tunneling spectroscopy.

7   Observability of the Mott-Insulating Phase

In this section we discuss various factors which might influence the observability of the Mott-insulating phase in realistic systems. An exponential increase of the resistivity of half-filled metallic SWNTs at low temperatures can also occur due to deformation-induced gaps Δ_d in the single-particle spectrum [30]. These gaps are in the range $\Delta_d < 0.02$ eV for typical SWNTs with $R \simeq 0.7$ nm. This value is comparable with (but somewhat smaller than) our estimate for the Mott gaps. We therefore cannot exclude the interplay between the two mechanisms suppressing electron transport at half-filling. Note that the deformation induced gaps decrease as $1/R^2$ with increasing nanotube radius [30], which should be contrasted to $1/R^{5/4}$ dependence for the Mott gaps. Moreover, deformation-induced gaps strongly depend on the chiral angle of SWNTs, whereas Mott gaps are determined by the radius of SWNTs and depend weakly on their chirality.

Isolated metallic SWNTs are half-filled systems. However, in generic experimental layouts the nanotubes are contacted to metallic electrodes. The difference ΔW in the work functions of a metal (typically Au or Pt) and nanotube results in a charge transfer between the nanotube and electrodes, which shifts the Fermi level of the nanotube downwards away from half-filling (for $\Delta W > 0$).

To be specific, consider a metallic SWNT surrounded by a coaxial cylindrical gate electrode of radius $R_s \gg R$. The nanotube ($x > 0$) contacts the yz-plane of metallic electrode ($x < 0$) at $x = 0$ (see inset of Fig. 4). The interaction term of the Hamiltonian (10) is modified by taking the electrostatic potential V_g of the gate as well as the image charge at the metallic electrode into account,

$$H_{int} = \int_0^\infty dx dx' \{ \frac{1}{2} \rho(x) [V(x - x') - V(x + x')] \rho(x') \tag{42}$$

$$+ \rho(x) M(x - x') e V_g sign(x') \} + \Delta W \int_0^\infty dx \rho(x),$$

The Fourier images of the kernels $V(x)$, $M(x)$ are given by,

$$\bar{V}_q = \frac{2e^2}{\kappa} \left\{ I_0(qR) K_0(qR) - I_0^2(qR) \frac{K_0(qR_s)}{I_0(qR_s)} \right\}, \quad \bar{M}_q = \frac{I_0(qR)}{I_0(qR_s)}, \tag{43}$$

with the modified Bessel functions I_0, K_0. The kernel $\bar{V}_q$ describes the long-range Coulomb interaction, $V(x) = 1/\kappa|x|$, for $R \ll |x| \ll R_s$. The interaction is screened at large distances $|x| \gg R_s$, so that $\bar{V}_{q=0} = (2e^2/\kappa) \ln(R_s/R)$, in agreement with our previous estimate (see text below Eq. (3)).

Minimization of the total energy given by the Hamiltonian (10) with the interaction term (42) determines the profile of the charge density $e\rho(x)$ and the deviation $-\Delta E(x)$ of the Fermi level from half-filling, $\Delta E = \rho/\nu$ (see Refs. [31,32] for details),

$$\Delta E(x) = \frac{\Delta W}{\nu \bar{V}(0)} \frac{\ln(R_s/R)}{\ln(z/R)} - \frac{ce V_g}{\nu \bar{V}(0)} \frac{x}{R_s}, \quad \text{for } R \ll x \ll R_s, \tag{44}$$

$$\Delta E(x) = \frac{\Delta W - e V_g}{1 + \nu \bar{V}(0)}, \quad \text{for } R_s \ll x, \tag{45}$$

where $c = (1/\pi) \int dx |I_0(x)|^{-1} \simeq 1.33$ and the Coulomb interaction is supposed to be strong, $\nu \bar{V}(0) \gg \ln(R_s/R)$.

Equation (44) shows that the density of charge transferred to SWNT due to the mismatch ΔW of the work-functions decays logarithmically slow with the distance x from the metallic electrode [31]. This is related to the poor screening of the Coulomb interaction in 1D SWNTs. The influence of a gate voltage on the charge density in SWNT is suppressed by a factor x/R_s in the vicinity of electrode, $x < R_s$, cf. Eqs. (44), (45). In order to achieve half-filling condition $\Delta E = 0$ in samples with small distances $d < R_s$ between the electrodes, one should compensate for this effect by applying a higher gate voltage.

The coordinate dependence of the charge density (or the deviation $\Delta E(x)$ from half-filling) is shown in Fig. 4 for several values of the gate voltage and the screening radius. If a substantial positive voltage is applied to the gate, the charge density in SWNT can change sign (from positive to negative) with increasing distance from the metallic electrode. Let us note that the Mott gap should result in the formation of half-filled incompressible regions near such "charge neutrality points", $\Delta E(x) = 0$. This is in contrast to isolated SWNTs where the Mott transition occurs uniformly in the whole system. The electron transport through the incompressible regions seems to be an interesting problem for future research.

8   Conclusions

We have investigated manifestations of electron correlations in metallic single-wall carbon nanotubes. Effective low-energy model of metallic SWNTs with arbitrary chirality is developed starting from microscopic considerations. The unrenormalized parameters of the model show very weak dependence on the chiral angle, which makes the low-energy properties of metallic SWNTs chirality-independent. In the absence of inter-branch electron scattering, the model corresponds to a two-channel Luttinger liquid. The impurity scattering in such Luttinger liquid is investigated using the memory function technique and the renormalization group method. We show that the intra-valley and inter-valley backscattering can not coexist at low energies. The properties of clean SWNTs are investigated beyond the Luttinger liquid limit. The ground state away from half-filling is found to be $4q_F$ spin density wave. At half-filling, the umklapp scattering affects the strongly interacting mode of total charge excitations. This makes the umklapp scattering a strongly relevant perturbation which gives rise to the Mott-insulating ground state. The latter is characterized by "binding" of electrons at two atomic sublattices of graphite. The energy gaps in all modes of excitations are evaluated within a self-consistent harmonic approximation. Finally, the observability of the Mott phase in realistic SWNTs is discussed.

9   Acknowledgments

The authors would like to thank B.L. Altshuler, G.E.W. Bauer, C. Dekker, Yu.V. Nazarov, S.Tans, S. Tarucha, Y. Tokura, and N. Wingreen for stimulating discussions. Financial support of the Royal Dutch Academy of Sciences (KNAW) is gratefully acknowledged.

References

1. For a recent review see C. Dekker, Physics Today **5**, 22 (1999)
2. S.J. Tans, M. H. Devoret, H. Dai, A. Thess, R. E. Smalley, L. J. Geerligs, C. Dekker, Nature (London) **386**, 474, (1997)
3. M. Bockrath, D. H. Cobden, B. L. McEuen, N. G. Chopra, A. Zettl, A. Thess, R. E. Smalley, Science, **275**, 1922 (1997)
4. J.W.G. Wildöer, L.C. Venema, A.G. Rinzler, R.E. Smalley, C. Dekker, Nature (London) **391**, 59 (1998)
5. T.W. Odom, J. Huang, P. Kim, C.M. Lieber, Nature (London) **391**, 62 (1998)
6. W. Clauss, D. J. Bergeron, A. T. Johnson, Phys. Rev. B **58**, R4266 (1998)

7. L.C. Venema, J.W.G. Wildoer, J.W. Janssen, S.J. Tans, H. Tuinstra, L.P. Kouwen-hoven, C. Dekker, Science **283**, 52 (1999)
8. M. Bockrath, D.H. Cobden, J. Lu, A.G. Rinzler, R.E. Smalley, L. Balents, and P.L. McEuen, Nature (London) **397**, 598 (1998)
9. C. Schonënberger, A. Bachtold, C. Strunk, J.-P. Salvetat, L. Forro, Appl. Phys. A 69, **283** (1999)
10. Z. Yao, H. Postma, L. Balents, and C. Dekker, to be published in Nature (London)
11. M.S. Fuhrer, J. Nygard, L. Shih, M. Bockrath, A. Zettl, and P. McEuen, submitted to Nature (London)
12. S.J. Tans, A.R.M. Verschueren, and C. Dekker, Nature (London) **393**, 49 (1998)
13. C. Kane, L. Balents and M. P. A. Fisher, Phys. Rev. Lett. **79**, 5086 (1997)
14. R. Egger and A. O. Gogolin, Phys. Rev. Lett. **79**, 5082 (1997); Eur. Phys. J. B **3**, 281 (1998)
15. H. Yoshioka and A.A. Odintsov, Phys. Rev. Lett., **82**, 374 (1999)
16. H. Yoshioka, preprint cond-mat/9903342
17. A.A. Odintsov and H. Yoshioka, Phys. Rev. B **59**, R10457 (1999)
18. P. R. Wallace, Phys. Rev. **71**, 622 (1947)
19. see e.g. E. Moore, B. Gherman, and D. Yaron, J. Chem. Phys. **106**, 4216 (1997)
20. H. Lin, L. Balents, and M.P.A. Fisher, Phys. Rev. B **58**, 1794 (1998)
21. T. Ando and T. Nakanishi, J. Phys. Soc. Jpn. **67**, 1704 (1998)
22. H. Mori, Prog. Theor. Phys. **33**, 423 (1965) **34**, 399 (1965); W. Götze and P. Wölfle, Phys. Rev. B **6**, 1126 (1972)
23. T. Giamarchi and H. J. Schulz, Phys. Rev. B **37**, 325 (1988)
24. V. L. Berezinskiǐ, Sov. Phys. JETP. **38**, 620 (1974)
25. N. Nagaosa, Solid State Communications **94**, 495 (1995)
26. H. J. Schulz, Phys. Rev. B **53**, R2959 (1996)
27. L. Balents and M. P. A. Fisher, Phys. Rev. B **55**, R11973 (1997)
28. Yu. A. Krotov, D.-H. Lee, and Steven G. Louie, Phys. Rev. Lett. **78**, 4245 (1997)
29. R.P. Feynman, *Statistical mechanics*, Reading, Mass., W.A. Benjamin, Inc., 1972
30. C.L. Kane and E.J. Mele, Phys. Rev. Lett. **78**, 1932 (1997)
31. A.A. Odintsov and Y. Tokura, to be published in *Proceedings of the LT-22, Helsinki, 1999*, preprint cond-mat/9906269
32. A.A. Odintsov and Y. Tokura, to be published.

Bosonization Theory of the Resonant Raman Spectra of Quantum Wires

Maura Sassetti[1] and Bernhard Kramer[2]

[1] Dipartamento di Fisica, INFM, Università di Genova, Via Dodecaneso 33
I-16146 Genova.
[2] I. Institut für Theoretische Physik, Universität Hamburg, Jungiusstraße 9
D-20355 Hamburg.

Abstract. We develop a Bosonization theory for the differential cross section for resonant Raman scattering on interacting electrons in quantum wires. The charge and spin density excitations, observed in recent Raman experiments, are identified. Near resonance, the hitherto unexplained "single particle excitations" are shown to originate in higher order collective spin excitations that are dressed with charge modes. A new selection rule for the inter-subband "single particle excitations" is predicted. Non-analytic power-law dependencies on photon energy and/or temperature of the intensities of the peaks in the resonant Raman spectra are derived which reflect the strength of the electron interaction.

1   Introduction

Interacting electrons, restricted to one spatial dimension, show certain paradigmatic features which are of extraordinary importance for modern solid state physics. The Hamiltonian of the interacting particles can be exactly diagonalized in terms of Bosonic modes within the Tomonaga-Luttinger liquid model [1–6]. This implies that the lowest-energy elementary excitations in one dimension are *collective*. Quasi-particles have a vanishingly small life time and the Fermi liquid model breaks down [7]. The Tomonaga-Luttinger liquid acquires fundamental importance especially in view of the considerable interest in quasi-one dimensional metallic molecules [4,5,8], recent developments in HTC-superconductivity [9], the possibilities to fabricate and experimentally investigate systematically electrical quantum transport in sub-micron semiconductor quantum wires [10,11] and carbon nanotubes [12]. Recently, even in the at the first glance somewhat remote area of the quantum Hall effect, due to a mapping of the edge states in the fractional quantum Hall regime to a so-called chiral Tomonaga-Luttinger system [13–15], the model turns out to gain great significance. All of these systems cannot be understood within Landau's phenomenological quasi-particle model and need fresh theoretical approaches. Unfortunately, up to now, straightforward and consistent experimental evidence for the validity of the model for quantum wires is still lacking, in spite of many efforts during the past decade [10–12,16–19].

A direct, and very powerful method to investigate the elementary excitations of condensed matter systems is inelastic scattering of light [20–22]. Especially in recent years, Raman scattering has been widely applied to sub-micron semiconductor structures [23–28]. Also, methods with high spatial resolution have been developed. They open new perspectives in investigating devices with spatial dimensions in the submicron region. A wealth of Raman scattering data on semiconductor quantum wires has been collected. They clearly show the existence of low-frequency collective excitations of the interacting electrons which can only be understood by taking into account the Coulomb interaction. Depending on whether the incoming and the scattered light are parallel or perpendicularly polarized, charge and spin density excitations (CDE and SDE) have been detected, respectively, when the photon energy $\hbar\omega_I$ is much larger than the fundamental energy gap of the semiconductor, E_G, i. e. "off-resonance".

When approaching resonance, i. e. $\hbar\omega_I \to E_G$, resonant structures are observed in the Raman spectra [23–27]. They are nearly independent of the polarization and exhibit approximately the dispersion relation of the non-interacting electrons. Therefore, they have been attributed to "single particle excitations" ("SPE"). Recently, it has been found, by using the Bosonization approach of the Tomonaga-Luttinger liquid model, that the "SPE" in the case of the (energetically lowest) intra-subband modes are physically due to higher-order spin density excitations that appear as peaks in the Raman spectra near resonance in both polarizations of incident and scattered light [29].

These findings suggest that the physics of the interacting one dimensional electron system can indeed be understood in terms of *collective* excitations that are dominated by interactions. For instance, one of the most important — though not unique — predictions of the theory of the Tomonaga-Luttinger liquid for a single occupied subband in a quantum wire, namely the *spin-charge separation*, are clearly observed in the Raman spectra. The *intraband* CDE are found to have the dispersion

$$\omega_\rho(q) = v_\rho(q)|q|, \tag{1}$$

with the velocity of propagation,

$$v_\rho(q) = v_F \left[1 + \frac{2V(q)}{\pi\hbar v_F}\right]^{-1/2}, \tag{2}$$

given by the Fermi velocity v_F renormalized by the Fourier transform of the potential of the Coulomb interaction, $V(q)$ [30]. On the other hand, the *intraband* SDE propagate approximately with the Fermi velocity since the exchange interaction is very small as compared to the Coulomb interaction,

$$\omega_\sigma(q) = v_F|q|. \tag{3}$$

This is completely consistent with the predictions of the Tomonaga-Luttinger model and suggests that the electron gas in a quantum wire might be candidate for genuine non-Fermi liquid behavior.

In order to confirm this surmise in more detail, it is necessary to investigate correlation functions in comparison with experiments and to demonstrate that they show the typical interaction-dominated power law behaviors predicted by the Tomonaga-Luttinger liquid model [5,8]. We will show below that this is indeed the case. At zero temperature, the strengths of the SDE peaks in the resonant Raman spectra that are related to higher order spin density correlation functions and which are usually denoted as "SPE" can be written in terms of integrals of the type

$$\int_{-\infty}^{\infty} \mathrm{d}x \, \frac{\exp{(iQx)}}{(1 + q_c^2 x^2)^{\mu(g)}} \tag{4}$$

with a "renormalized photon wave number"

$$Q \equiv \frac{\hbar\omega_I - E_G - \hbar v_F q/2}{\hbar v_F}. \tag{5}$$

The cutoff q_c reflects the range of the interaction in wave number space. For Coulomb interaction, q_c is of the order of the inverse diameter of the quantum wire, d. For quantitative purposes, one can assume $q_c \approx k_F$ [31]. The exponent

$$\mu(g) = \frac{g + g^{-1}}{8} - \frac{1}{4} \tag{6}$$

contains the interaction parameter

$$g \equiv \frac{v_F}{v_\rho(q \to 0)}. \tag{7}$$

The exponent $\mu(g)$ is characteristic of Tomonaga-Luttinger liquid correlation functions [5,8].

The results for the Raman cross section corresponding to the intraband excitations can be reformulated to include also the interband collective modes by using approximations which are very similar in spirit to those employed in the Tomonaga-Luttinger liquid model [32]. We will show that such a generalization leads to predicting a strikingly novel "selection rule", namely that only inter-subband SDE with positive group velocity can appear as "SPE" when approaching resonance [33]. It should be extremely easy to verify or falsify experimentally this prediction with presently available technology, thus providing another indication for the predictive power of the Bosonization method for inelastic light scattering on semiconductor quantum wires.

We claim that by measuring on quantum wires the dependence of the intensity of the so-called "SPE" feature on the energy of the incident photons when approaching resonance, insight can be gained into whether or not non-Fermi liquid behavior dominates the collective low energy excitations of the one dimensional electron gas.

The paper is organized as follows: in section 2, we describe briefly the possible electronic pair excitations in a two-subband quantum wire. The linearized two-band Hamiltonian is presented in section 3 and the corresponding eigenmodes in section 4. Section 5 contains an introduction to the theory of Raman scattering, especially in view of its application to quantum wires. The cross sections for resonant intra-subband and inter-subband scattering are given in the sections 6 and 7, respectively. Section 8 compares the results with experimental data and conclusions are drawn in the final section 9.

2   Pair Excitations of Non-interacting Electrons in One Dimension

We consider quasi-one dimensional confined non-interacting electrons with an effective mass m. The corresponding one-particle energy spectrum consists of parabolic subbands,

$$\varepsilon_j(k) = \varepsilon_j + \frac{\hbar^2 k^2}{2m}, \tag{8}$$

where ε_j $(j = 1, 2, 3, \dots)$ are the confinement energies and k is the wave number. We assume that the Fermi energy E_F is such that only two subbands $(j = 1, 2)$ are occupied and the corresponding Fermi velocities $v_{Fj} = [2(E_F - \varepsilon_j)/m]^{1/2}$ are roughly the same. This can approximately be achieved in experiment. For example, for a box-like confinement potential, the third subband will be energetically far away from the second one and it is possible to adjust E_F well in between. Even when the subbands are energetically almost equidistant — the case of ideal parabolic confinement realized in many of the experiments [34] — the available experimental results suggest that the observed excitations at low frequencies are dominated by transitions within and between the energetically lowest bands. Possible pair excitations of such a non-interacting electron system are qualitatively shown in Fig. 1.

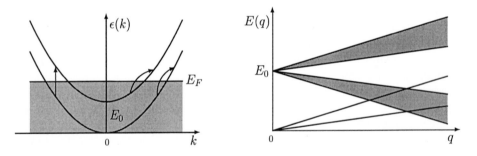

Fig. 1. Left: One-electron subbands of a quantum wire with spacing $\varepsilon_2 - \varepsilon_1 \equiv E_0$ and Fermi energy E_F such that two subbands are partially filled with particles. Arrows indicate possible pair excitations. Right: pair excitation energies of the non-interacting quasi-one dimensional electron gas linearized for small wave numbers.

3   The Linearized Two-Band Hamiltonian

In order to describe the collective electronic excitations in a quantum wire, we consider the Hamiltonian of non-interacting electrons in two subbands $(j = 1, 2)$

$$H_0 = \sum_{jsk} \varepsilon_j(k) c^{\dagger}_{js}(k) c_{js}(k), \tag{9}$$

with $c_{js}^\dagger(k)$, $c_{js}(k)$ the Fermion operators corresponding to subband j, spin quantum number s and wave number k.

The interaction energy

$$H_{int} = \sum_{ss'} \sum_{ijlm} \sum_{qkk'} V_{ijlm}(q) c_{is}^\dagger(k+q) c_{js'}^\dagger(k') c_{ls'}(k'+q) c_{ms}(k) \qquad (10)$$

contains the matrix elements $V_{ijlm}(q)$ that are obtained by projecting the three dimensional screened Coulomb potential

$$V(r) = \frac{e^2}{4\pi\epsilon r} e^{-\alpha r} \equiv \frac{U_0}{r} e^{-\alpha r} \qquad (11)$$

onto the subbands and Fourier transforming with respect to x, the coordinate in the direction of the quantum wire [31].

First, we need to write the total Hamiltonian $H = H_0 + H_{int}$ as a quadratic form in the densities

$$\rho_{ij,s}(q) \equiv \sum_k c_{is}^\dagger(k+q) c_{js}(k) \equiv \rho_{ij,s}^{(-)}(q) + \rho_{ij,s}^{(+)}(q) . \qquad (12)$$

Here, $\rho^{(+)}$ and $\rho^{(-)}$ correspond to the branches of the spectrum with positive and negative velocities, respectively. With these, intra- and interband charge and spin densities in branches λ may be defined,

$$\rho_{ij}^{(\lambda)}(q) \equiv \frac{1}{\sqrt{2}} \left[\rho_{ij,+}^{(\lambda)}(q) + \rho_{ij,-}^{(\lambda)}(q) \right] ,$$
$$\sigma_{ij}^{(\lambda)}(q) \equiv \frac{1}{\sqrt{2}} \left[\rho_{ij,+}^{(\lambda)}(q) - \rho_{ij,-}^{(\lambda)}(q) \right] . \qquad (13)$$

Apart from a Bogolubov transformation, they represent the eigenmodes of the Hamiltonian, as will be seen below.

Of crucial importance for the construction of the eigenmodes is to generate commutation relations of the Hamiltonian with the densities of the form

$$\left[H_0, \rho_{ij,s}^{(\lambda)}(q) \right] \propto \rho_{ij,s}^{(\lambda)}(q) . \qquad (14)$$

From (9) and (12) one obtains

$$[H_0, \rho_{ij,s}^{(\lambda)}(q)] = (\varepsilon_i - \varepsilon_j) \rho_{ij,s}^{(\lambda)}(q) + \frac{\hbar^2 q}{m} \sum_{k,\lambda=\pm} k c_{is}^{(\lambda)\dagger}(k+q) c_{js}^{(\lambda)}(k) + O(q^2) . \qquad (15)$$

For the intraband densities, $i = j$, we obtain the required commutator if we assume that in the sum on the right hand side $k \approx \lambda k_{Fi}$, consistent with the linearization of the dispersion [3]. For $i \neq j$, we replace $k \approx \lambda(k_{F1} + k_{F2})/2$ consistent with $\varepsilon_2 - \varepsilon_1 \equiv E_0 \ll E_F$. This corresponds to linearizing the two subbands around the mean Fermi wave number $k_F \equiv (k_{F1} + k_{F2})/2$.

The commutators of the densities are

$$[\rho_{12,s}^{(\lambda)}(q), \rho_{ii,s'}^{(\lambda')}(q')] = (-1)^i \delta_{\lambda\lambda'} \delta_{ss'} \rho_{12,s}^{(\lambda)}(q+q'), \tag{16}$$

$$[\rho_{12,s}^{(\lambda)}(q), \rho_{21,s'}^{(\lambda')}(-q')] = \sum_k \Big[c_{1s}^{(\lambda)\dagger}(k+q-q') c_{1s}^{(\lambda)}(k)$$
$$- c_{2s}^{(\lambda)\dagger}(k-q') c_{2s}^{(\lambda)}(k-q) \Big] \delta_{\lambda\lambda'} \delta_{ss'}. \tag{17}$$

Relations (16) and (17) imply that *interband* and *intraband* modes are *not* decoupled. However, they may be decoupled when considering expectation values in the ground state, and by assuming

$$\langle \rho_{12,s}^{(\lambda)} \rangle = 0, \tag{18}$$

and

$$\langle c_{is}^{(\lambda)\dagger}(k) c_{is}^{(\lambda)}(k') \rangle = \delta_{k,k'} n_{is}^{(\lambda)}(k) \tag{19}$$

where $n_{is}^{(\lambda)}(k)$ is the Fermion particle number. One can then show that the right hand side of (17) gives $\delta_{\lambda\lambda'} \delta_{ss'} \delta_{q,q'} (L/2\pi)(k_{F1} - k_{F2} - \lambda q)$ (L system length), similar as in the one-band case. Commutator (17) has now the required form necessary for formally describing the interband excitations as Bosons [3].

With this, the free part of the Hamiltonian can be decomposed as

$$H_0 = H_{0\,\text{intra}} + H_{0\,\text{inter}}, \tag{20}$$

with the contribution corresponding to intra-subband excitations

$$H_{0\,\text{intra}} = \frac{\hbar\pi}{L} \sum_{i=1,2} v_{Fi} \sum_{q\lambda s} \rho_{ii,s}^{(\lambda)}(q) \rho_{ii,s}^{(\lambda)}(-q), \tag{21}$$

and the inter-subband part

$$H_{0\,\text{inter}} = \frac{2\pi\hbar v_F}{L} \sum_{q\lambda s} \rho_{12,s}^{(\lambda)}(q) \rho_{21,s}^{(\lambda)}(-q), \tag{22}$$

where v_F is the velocity corresponding to the mean Fermi wave number k_F.

4   The collective eigenmodes

With the above assumptions (18) and (19), the total Hamiltonian, including matrix elements of the interaction (10), can now be written in analogy to (20) as a sum of two contributions

$$H = H_{\text{intra}} + H_{\text{inter}} \tag{23}$$

which describe decoupled *intraband* and *interband* collective excitations [32]. The corresponding frequency spectra can be determined exactly but eventually have to be evaluated numerically. One obtains four *intraband* modes [29],

$$\omega_{\nu 0}^{\pm} = v_{\nu 0}^{\pm}(q)|q|,\tag{24}$$

with $\nu = \rho, \sigma$ for the CDE and SDE, respectively, $\pm$ denoting the in-phase and out-of-phase branches. For $q \to 0$, the velocities are given in leading order in the Coulomb interaction by

$$v_{\rho 0}^{+}(q) = \sqrt{4v_0(q)\,(\tilde{v}_{F1} + \tilde{v}_{F2} + 2v_{ex0})},\tag{25}$$

$$v_{\rho 0}^{-}(q) = \sqrt{\tilde{v}_{F1}\tilde{v}_{F2}}\,\sqrt{\frac{2(v_{F1}+v_{F2}) - (\tilde{v}_{F1}+\tilde{v}_{F2}) + 2u_0}{\tilde{v}_{F1} + \tilde{v}_{F2} + 2v_{ex0}}},\tag{26}$$

$$v_{\sigma 0}^{+}(q) = \sqrt{\tilde{v}_{F1}(2v_{F1} - \tilde{v}_{F1}) - \frac{2v_{F1}v_{ex0}^2}{v_{F1}+v_{F2}}},\tag{27}$$

$$v_{\sigma 0}^{-}(q) = \sqrt{\tilde{v}_{F2}(2v_{F2} - \tilde{v}_{F2}) - \frac{2v_{F2}v_{ex0}^2}{v_{F1}+v_{F2}}}.\tag{28}$$

The Fourier transformed long range part of the interaction potential, projected onto a given subband, $V(q)$, appears solely in the in-phase charge density mode [29] as the wave number dependent velocity

$$v_0(q) = 2u_0|\ln(qd)|\tag{29}$$

All of the other modes feel only the short range parts of the interaction and we abbreviated the prefactor of the interaction, cf. (11), by $U_0/h = u_0$. For convenience, we have defined here velocities renormalized by the interaction,

$$\tilde{v}_{Fj} = v_{Fj} + \frac{V_{jjjj}(2k_{Fj})}{h}\tag{30}$$

and

$$v_{ex0} = \frac{V_{1221}(k_{F1} + k_{F2})}{h}.\tag{31}$$

In the following, we will also need the velocity representing the inter-subband exchange interaction,

$$v_{ex1} = \frac{V_{1212}(k_{F1} + k_{F2})}{h},\tag{32}$$

and the velocity representing the inter-subband Coulomb term,

$$v_1 = \frac{2V_{1212}(0)}{h}.\tag{33}$$

For the four interband modes, we obtain after lengthy but straightforward calculations in the limit of small q [32]

$$\omega_{\nu 1}^{\pm}(q) = \omega_{\nu 1}^{\pm}(0)\left(1 \pm A_{\nu 1}^{\pm}\frac{q^2 v_{\mathrm{F}}^2}{\omega_0^2}\right). \tag{34}$$

Here, the frequency scales are given by

$$\omega_{\rho 1}^{\pm}(0) = \omega_0\sqrt{1 + (1 \pm 1)\frac{2v_1}{v_{\mathrm{F}}} \mp \frac{2v_{\mathrm{ex1}}}{v_{\mathrm{F}}}}, \tag{35}$$

$$\omega_{\sigma 1}^{\pm}(0) = \omega_0\sqrt{1 \mp \frac{2v_{\mathrm{ex1}}}{v_{\mathrm{F}}}}, \tag{36}$$

with $\omega_0 = E_0/\hbar$ and the constants

$$A_{\rho 1}^{\pm} = \frac{v_{\mathrm{F}}}{v_1}\left[1 + (4 \pm 1)\frac{v_1}{v_{\mathrm{F}}} + 2(1 \pm 1)\left(\frac{v_1}{v_{\mathrm{F}}}\right)^2\right],$$

$$A_{\sigma 1}^{\pm} = -\frac{v_{\mathrm{F}}}{v_{\mathrm{ex1}}}.$$

The above expressions are valid to the order $O((v_{\mathrm{ex1}}/v_{\mathrm{F}})^2)$. The excitation spectra (24) and (34) are shown schematically in Fig. 2. We wish to emphasize here again that they correspond to collective modes.

For small wave numbers, the four energetically lowest branches of the spectrum that start at zero frequency for $q \to 0$ correspond to the *intraband* charge and spin density modes, $\mathrm{CDE}_0^{\pm}$ and $\mathrm{SDE}_0^{\pm}$, respectively. The two lowest branches are the SDE. They are linear in the wave number with slopes approximately equal to the Fermi velocities $v_{\mathrm{F}1,2}$ since the exchange interaction is very small as compared with the Fermi velocities. The two next-higher branches are the charge density modes. The uppermost intraband CDE, symmetric in the charge densities of the two subbands, reflects the Fourier transform of the interaction $v_0(q)$ [7,29]. The slope of the energetically lower (anti-symmetric) intraband-CDE is proportional the geometrical average of the two Fermi velocities.

The four modes which start at $q = 0$ with finite frequencies are *interband* charge and spin excitations. The energetically highest branch with the positive group velocity corresponds to a symmetric interband charge density mode $\approx \rho_{12}^{(+)} + \rho_{12}^{(-)}$. It is strongly shifted to higher energy by the Coulomb repulsion (depolarization shift). The anti-symmetric branch, $\approx \rho_{12}^{(+)} - \rho_{12}^{(-)}$, does not feel the Coulomb repulsion at small wave numbers and has a negative group velocity. It is degenerate with the anti-symmetric interband SDE at an energy which is given approximately by the interband spacing plus the exchange energy, since the latter is very small. For typical state-of-the-art AlGaAs/GaAs quantum wires $E_0 = 3$ meV and $E_0 v_{\mathrm{ex1}}/v_{\mathrm{F}} \approx 0.2$ meV [27]. The two remaining branches of the spectrum correspond to the symmetric and anti-symmetric SDE ($\approx \sigma_{12}^{(+)} \pm \sigma_{12}^{(-)}$), starting at $\omega_{\sigma 1}^{\pm}(0) \approx \omega_0(1 \mp v_{\mathrm{ex1}}/v_{\mathrm{F}})$, respectively.

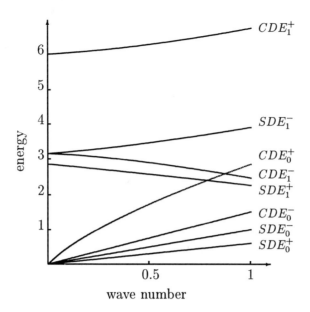

Fig. 2. Schematic energy spectrum of the intra- and inter-subband charge and spin density modes of a two-band quantum wire for small wave number. Energy in meV; wave numbers in $10^5\,\mathrm{cm}^{-1}$. Parameters correspond approximately to state-of-the-art AlGaAs/GaAs quantum wires: $E_0 = 3\,\mathrm{meV}$, $v_1/v_F = 3/4$, $v_{ex1}/v_F = 1/15$, $k_{F1} = 10^6\,\mathrm{cm}^{-1}$, $k_{F2} = 0.6\,k_{F1}$.

Having presented a detailed overview over the spectral features of the two-band model of the quantum wire, we will proceed in the following by evaluating the Raman cross section by using explicitly the Bosonization transformation which leads to the above eigenmodes.

5   The Raman Cross Section

Within the standard theory of Raman scattering [20–22], the differential cross section is given by

$$\frac{d\sigma}{d\Omega d\omega} = \left(\frac{e^2}{m_0 c^2}\right)^2 \frac{\omega_O}{\omega_I} \frac{n(\omega)+1}{\pi}\, \mathcal{I}\mathrm{m}\chi(\boldsymbol{q},\omega)\,, \tag{37}$$

(m_0 bare electron mass) with the energy and momentum transfers $\hbar\omega = \hbar\omega_I - \hbar\omega_O$ and $\boldsymbol{q} = \boldsymbol{k}_I - \boldsymbol{k}_O$, respectively, and the Bose distribution $n(\omega)$.

The correlation function

$$\chi(\boldsymbol{q},t) = i\Theta(t)\langle[N^\dagger(\boldsymbol{q},t), N(\boldsymbol{q},0)]\rangle\,, \tag{38}$$

contains the generalized density operator $N(\boldsymbol{q})$ which is of the form

$$N(\boldsymbol{q}) = \sum_{\alpha\alpha'} \gamma_{\alpha\alpha'}(\boldsymbol{k}_{\mathrm{I}}, \boldsymbol{k}_{\mathrm{O}}) c_{\alpha}^{\dagger} c_{\alpha'} \,. \tag{39}$$

It contains the creation and annihilation operators of the electrons in the conduction band states $|\alpha\rangle$, $|\alpha'\rangle$. The transition matrix elements $\gamma_{\alpha\alpha'}$ are

$$\gamma_{\alpha\alpha'} = \langle\alpha|e^{i\boldsymbol{q}\cdot\boldsymbol{r}}|\alpha'\rangle\,(\boldsymbol{e}_{\mathrm{I}}\cdot\boldsymbol{e}_{\mathrm{O}})$$

$$+ \frac{1}{m_0} \sum_{\beta} \left[\frac{\langle\alpha|J_{\mathrm{O}}|\beta\rangle\langle\beta|J_{\mathrm{I}}|\alpha'\rangle}{\varepsilon_{\alpha'} - \varepsilon_{\beta} + \hbar\omega_{\mathrm{I}}} + \frac{\langle\alpha|J_{\mathrm{I}}|\beta\rangle\langle\beta|J_{\mathrm{O}}|\alpha'\rangle}{\varepsilon_{\alpha} - \varepsilon_{\beta} - \hbar\omega_{\mathrm{I}}} \right] , \tag{40}$$

where $|\beta\rangle$ denote states in the valence bands, $\boldsymbol{e}_{\mathrm{I}}, \boldsymbol{e}_{\mathrm{O}}$ the polarization vectors of incoming and outgoing light and

$$J_{\mu} = (\boldsymbol{e}_{\mu}\cdot\boldsymbol{\pi})\,e^{\pm i\boldsymbol{k}_{\mu}\cdot\boldsymbol{r}} , \tag{41}$$

$(\mu = \mathrm{I}, \mathrm{O})$, the projections of current operator to the direction of the polarization with "$+$" appearing in the exponent when $\mu = \mathrm{I}$. The operator

$$\boldsymbol{\pi} = \boldsymbol{p} + \frac{1}{4m_0 c^2}\,(\hbar\boldsymbol{\sigma}\times\boldsymbol{\nabla}V) \tag{42}$$

represents the momentum. It contains the spin operator $\boldsymbol{\sigma}$, which has the Pauli matrices as components, and the gradient of the potential energy, $\boldsymbol{\nabla}V$, in the spin-orbit contribution. The latter is necessary in order to excite the SDE. The first term in (40) results from the $\boldsymbol{A}^2$-part of the Hamiltonian, while the second and third terms are due to the terms $\propto (\boldsymbol{\pi}\cdot\boldsymbol{A})$ ($\boldsymbol{A}$ vector potential). Near resonance, $\hbar\omega_{\mathrm{I}} \approx E_{\mathrm{G}}$, the fundamental energy gap of the semiconductor, the *third* term dominates. We concentrate here on this latter contribution.

In order to proceed further, the states have to be specified. With the above two subbands, we have $|\alpha\rangle \equiv |jsk\rangle$ ($j = 1, 2$). The splitting of the valence quasi-one dimensional subbands is neglected. This is reasonable since the effective electron mass of the valence band is large as compared to that of the conduction band in quantum wires based on AlGaAs/GaAs hetero-structures. We decompose the states with spin quantum number s into a plane wave phase factor with the wave number k, and factor which is lattice periodic in the direction of the wire. Then, we get

$$N(q) = \sum_{ss'ijk} \frac{\gamma_{is\,js'}(\boldsymbol{k}_{\mathrm{I}}, \boldsymbol{k}_{\mathrm{O}})}{D_i(k, q)} c_{is}^{\dagger}(k + q) c_{js'}(k) , \tag{43}$$

with the Fermion operators $c_{is}^{\dagger}(k)$, $c_{js}(k)$ and q is the x-component of $\boldsymbol{k}_{\mathrm{I}} - \boldsymbol{k}_{\mathrm{O}}$. The transition matrix

$$\gamma_{is\,js'}(\boldsymbol{k}_{\mathrm{I}}, \boldsymbol{k}_{\mathrm{O}}) = \delta_{ss'}\,\boldsymbol{e}_{\mathrm{I}}\cdot\boldsymbol{\Gamma}_{\|ij}\cdot\boldsymbol{e}_{\mathrm{O}} + i(\boldsymbol{e}_{\mathrm{I}}\times\boldsymbol{e}_{\mathrm{O}})\cdot\boldsymbol{\Gamma}_{\perp ij}\cdot\boldsymbol{S}_{ss'} \tag{44}$$

describes the dependence on the relative polarizations of the incoming and the scattered light, $e_{I,O}$. The matrices $\Gamma_\parallel$ and $\Gamma_\perp$ consist of the transition amplitudes between the valence and the conduction bands. The vector $\boldsymbol{S}_{ss'}$ contains as components the matrix elements of the operator of the spin $\boldsymbol{\sigma}$ in the spinors corresponding to the electron states.

The first term in (44) represents the momentum contribution, $\boldsymbol{p}$, to J_μ. It will be seen to probe charge density excitations when assuming constant energy denominator. The second term originates in the presence of spin-orbit coupling in J_μ. It is associated with spin flips, and therefore probes spin density excitations if the energy denominator $D_i(k, q)$ is constant. The latter,

$$D_i(k, q) = \varepsilon_i(k + q) - \varepsilon_v(k + q - k_{Ix}) - \hbar\omega_I \,, \tag{45}$$

contains the difference between the energies of the valence and conduction bands ε_v and ε_i, respectively. It originates formally in the second perturbation theory that is employed when expanding the cross section consistently to the order $(\boldsymbol{\pi} \cdot \boldsymbol{A})^2$.

In the following, we are interested in general features of the cross section which are related to the collective excitations, rather than attempting a full calculation including all of the effects of the transition matrix elements. Therefore, we assume $\Gamma_{\parallel ij} = \gamma_{\parallel ij}\mathbf{I}$ and $\Gamma_{\perp ij} = \gamma_{\perp ij}\mathbf{I}$. Then

$$\gamma_{is\,js'}(\boldsymbol{k}_I, \boldsymbol{k}_O) = \delta_{ss'} \left[\gamma_{\parallel ij}\boldsymbol{e}_I \cdot \boldsymbol{e}_O + \gamma_{\perp ij}|\boldsymbol{e}_I \times \boldsymbol{e}_O|s \right] \,. \tag{46}$$

Here, we have also assumed a coordinate system such that $\mathbf{S}_{ss'} \parallel \hat{\boldsymbol{z}}$, and the polarization vectors perpendicular to $\hat{\boldsymbol{z}}$.

Given the Hamiltonian being diagonalized in terms of the Bosonic excitations, the cross section can be calculated by numerically evaluating the correlation function (38). This is in principle exactly possible within the present model. In practice, it is a considerable task which can only be performed with reasonable effort by using further approximations. In the following section, we first outline the example of the intraband SDE in the one-subband limit, in order to demonstrate the Bosonization technique. Then, we review the results for the interband case.

6   Resonant Intraband Raman Spectra

In this section we concentrate on the limit of only one occupied subband. Thus, we drop all unnecessary indices in order to facilitate the notations. First, we rewrite $N(q)$ by linearizing $D(k, q)$ in (43) around $\pm k_F$ consistent with the linearization of the electron dispersion when diagonalizing the Hamiltonian. We decompose the Fermionic operators into right and left moving branches,

$$c_s \equiv c_s^{(+)} + c_s^{(-)} \tag{47}$$

This gives

$$N(q) = \sum_{s\lambda k} \frac{\gamma_{\parallel} e_{\mathrm{I}} \cdot e_{\mathrm{O}} + is\gamma_{\perp}|e_{\mathrm{I}} \times e_{\mathrm{O}}|}{D^{(\lambda)}(k,q)} c_s^{(\lambda)\dagger}(k+q)c_s^{(\lambda)}(k), \qquad (48)$$

with

$$D^{(\lambda)}(k,q) = E_{\mathrm{G}} - \hbar\omega_{\mathrm{I}} + \hbar v_{\mathrm{F}}(\lambda k - k_{\mathrm{F}}) + \lambda\hbar v_{\mathrm{F}} q. \qquad (49)$$

Here, $E_{\mathrm{G}} = E_{\mathrm{G}}^0 + E_{\mathrm{F}}$ is the energy difference between the lowest conduction subband and the dispersionless valence band, E_{G}^0, plus the Fermi energy. The zero of energy has been chosen such that $\varepsilon_1 = 0$.

As in previous works [29,32], we start by expanding the denominator for large $\hbar\omega_{\mathrm{I}} - E_{\mathrm{G}}$. This yields in lowest order

$$N_0(q) = \frac{\sqrt{2}}{E_{\mathrm{G}} - \hbar\omega_{\mathrm{I}}} \sum_{\lambda} \left[\gamma_{\parallel} e_{\mathrm{I}} \cdot e_{\mathrm{O}} \rho^{(\lambda)}(q) + i\gamma_{\perp}|e_{\mathrm{I}} \times e_{\mathrm{O}}|\sigma^{(\lambda)}(q) \right], \qquad (50)$$

with $\rho^{(\lambda)}(q)$ and $\sigma^{(\lambda)}(q)$ defined in analogy to (13). This result predicts the "classical selection rule", namely that charge and spin excitations appear in the Raman spectra in parallel and perpendicular polarization of incident and scattered light, respectively, for photon frequencies far from the fundamental gap. The contribution to the correlation function corresponding to (50) is ($\omega > 0$)

$$\mathcal{I}m\chi_0(q,\omega) = \frac{Lq}{(E_{\mathrm{G}} - \hbar\omega)^2} \left[\frac{v_\rho(q)}{v_{\mathrm{F}}}(\gamma_{\parallel} e_{\mathrm{I}} \cdot e_{\mathrm{O}})^2 \delta(\omega - \omega_\rho(q)) \right.$$

$$\left. + (\gamma_{\perp}|e_{\mathrm{I}} \times e_{\mathrm{O}}|)^2 \delta(\omega - v_\sigma|q|) \right]. \qquad (51)$$

The next higher order contribution to the generalized density operator is

$$N_1(q) = -\frac{1}{2L} \frac{h v_{\mathrm{F}}}{(E_{\mathrm{G}} - \hbar\omega_{\mathrm{I}})^2} \times \sum_{\lambda k} \left\{ \gamma_{\parallel}(e_{\mathrm{I}} \cdot e_{\mathrm{O}}) \left[\rho^{(\lambda)}(k)\rho^{(\lambda)}(k-q) \right. \right.$$

$$\left. \left. + \sigma^{(\lambda)}(k)\sigma^{(\lambda)}(k-q) \right] + 2i\gamma_{\perp}|e_{\mathrm{I}} \times e_{\mathrm{O}}|\rho^{(\lambda)}(k)\sigma^{(\lambda)}(k-q) \right\}. \quad (52)$$

By using this in evaluating the correlation function (38), one predicts that due to the above term $\propto \sigma^{(\lambda)}(k)\sigma^{(\lambda)}(k-q)$ there will be a resonant contribution to the spectrum at the frequency of the SDE also in the parallel configuration. Its frequency and temperature dependence is given by

$$\mathcal{I}m\chi_1(q,\omega) = \frac{Lq}{12} \frac{(\hbar\gamma_{\parallel} v_\sigma)^2}{(E_{\mathrm{G}} - \hbar\omega_{\mathrm{I}})^4} \left[\left(\frac{\pi k_{\mathrm{B}} T}{\hbar v_\sigma} \right)^2 + \frac{q^2}{2} \right] \delta(\omega - v_\sigma|q|). \qquad (53)$$

On the other hand, a similar resonant structure due to charge modes violating the "classical selection rule", is absent in the perpendicular polarization since the combination of operators $\rho^{(\lambda)}(k)\sigma^{(\lambda)}(k-q)$ does not yield a monochromatic contribution to the correlation function (38) at the frequency of the CDE [29].

In order to obtain more detailed insight into the behavior closer to resonance, the energy denominator $D^{(\lambda)}(k,q)$ has to be taken into account without expanding, as given in (49). This can be done by using the Bosonization method. First, we rewrite the product of Fermion operators as

$$c_s^{(\lambda)\dagger}(k+q)c_s^{(\lambda)}(k) = \frac{1}{L}\int_{-\infty}^{\infty} dy e^{iy(k-\lambda k_F+q/2)} I_s^{(\lambda)}(q,y),\qquad (54)$$

with the integral

$$I_s^{(\lambda)}(q,y) = \int_{-\infty}^{\infty} dx e^{i(\lambda k_F y + qx)} \Psi_s^{(\lambda)\dagger}\left(x+\frac{y}{2}\right)\Psi_s^{(\lambda)}\left(x-\frac{y}{2}\right),\qquad (55)$$

which equivalently can be rewritten by using the Boson fields [3]

$$\Phi_s^{(\lambda)}(x,y) = -\frac{4\pi\lambda}{\sqrt{2L}}\sum_{q>0}\frac{\exp(i\lambda qx)}{q}\sin\left(\frac{qy}{2}\right)\left[\rho^{(\lambda)}(-\lambda q)+s\sigma^{(\lambda)}(-\lambda q)\right]\qquad (56)$$

into the form

$$I_s^{(\lambda)}(q,y) = \frac{i\lambda}{2\pi y}\int_{-\infty}^{\infty} dx e^{iqx} e^{i\Phi_s^{(\lambda)\dagger}(x,y)} e^{i\Phi_s^{(\lambda)}(x,y)}.\qquad (57)$$

In order to calculate the correlation function (38) one needs the Heisenberg operators corresponding to the charge and spin densities. They are obtained by expressing $\nu^{(\lambda)}$ ($\nu = \rho, \sigma$) in terms of the eigenmodes of the quadratic Hamiltonian for which the time evolution is given by the eigenenergies. For simplicity, we assume that q-dependence of the charge mode can be approximated by a *constant* sound velocity renormalized by the spatial average of the interaction, $v_\rho = v_F/g$ with the interaction parameter $g^{-2} = 1 + 2V(0)/\hbar\pi v_F$. This can be done for any finite range of the interaction potential and can be justified by observing *a posteriori* that for the interpretation of the experimental data the parameters are such that the small wave numbers contribute predominantly to the integrals. We extract the time dependencies that correspond to peak-like structures due to SDE when evaluating the integrals, namely

$$\chi(q,t) \propto e^{i\omega_\sigma(q)t},\qquad (58)$$

and neglect the exchange interaction matrix elements that would renormalize the velocity of the spin density mode v_σ with respect to v_F.

The final result, related to the SDE, can be written as

$$\mathcal{I}m\,\chi(q,\omega) = \left[\gamma_\parallel^2(e_I\cdot e_O)^2 I_\parallel + \gamma_\perp^2|e_I\times e_O|^2 I_\perp\right]\delta(\omega - v_\sigma q).\qquad (59)$$

The integrals representing the peak strength are given by

$$I_n(q,Q) = -\frac{2iL}{\pi}\int_0^{\infty} dy_1 \int_0^{\infty} dy_2 \int_{-\infty}^{\infty} dx \sin(qx)\cos[Q(y_1-y_2)]$$

$$\times F(y_1)F(y_2)\mathrm{Hyp}_n\left(Z(x,y_1,y_2)\right).\qquad (60)$$

Here, we have defined the functions $\mathrm{Hyp}_n(Z) = \cosh Z, \sinh Z$ depending on whether $n = \|$ or $n = \perp$, respectively. Furthermore, we have introduced

$$F(y) = \frac{1}{y(1 + q_c^2 y^2)^{\mu - 1/4}} \left[\frac{\hbar \beta v_\rho}{y} \sinh\left(\frac{\pi y}{\hbar \beta v_\rho}\right) \right]^{-2\mu - 1/2}$$

$$\times \left[\frac{\hbar \beta v_\sigma}{y} \sinh\left(\frac{\pi y}{\hbar \beta v_\sigma}\right) \right]^{-1/2}, \tag{61}$$

and

$$Z(x, y_1, y_2) = \frac{1}{2} \frac{J(x + y_+)J(x - y_+)}{J(x + y_-)J(x - y_-)}, \tag{62}$$

where

$$J(z) = \ln\left(1 - iq_c z\right) + \ln\left[\frac{\hbar \beta v_\sigma}{\pi z} \sinh\left(\frac{\pi z}{\hbar \beta v_\sigma}\right)\right]. \tag{63}$$

The variables are defined as $y_\pm \equiv (y_1 \pm y_2)/2$, $\beta \equiv (k_B T)^{-1}$, μ is given by (6) and the "reduced photon wave number" by (5).

For $Q \gg q$, one can expand $Z(x, y_1, y_2)$ for small y_1 and y_2. With this, the integral (60) can be evaluated in lowest order,

$$I_n \approx \frac{Lq}{(\hbar v_F)^2} \left[\frac{q^2}{24} + \left(\frac{\pi k_B T}{\hbar v_\sigma}\right)^2 \right]^p \left| \frac{d^p}{dQ^p} L_\mu(Q) \right|^2 \tag{64}$$

with $p = 1$ and 0 when $n = \|$ and $n = \perp$, respectively, and

$$L_\mu(Q) \equiv \int_0^\infty dy\, e^{iQy}\, y F(y), \tag{65}$$

with $F(y)$ defined in (61).

Two limiting cases are of particular interest with respect to experiments. On the one hand, for $Q \gg q_c$, which corresponds to photon energies that are considerably above E_G, one obtains

$$L_\mu(Q) \approx \frac{1}{iQ}, \tag{66}$$

independent of the interaction strength. This gives the following dependence of the peak strength on the frequency of incident photons,

$$I_n \propto \left(\frac{q_c}{Q}\right)^{2(p+1)}. \tag{67}$$

On the other hand, with $q \ll Q \ll q_c$ and for intermediate g, say $g > 0.2$,

$$L_\mu(Q) \approx \frac{\Gamma(1 - \mu)\Gamma(1/2 - \mu)}{2\sqrt{\pi} q_c} \left(\frac{2q_c}{iQ}\right)^{1 - 2\mu}. \tag{68}$$

In this parameter region, the strength of the interaction is clearly reflected in the power law of the dependence of the peak intensities on the "renormalized photon wave vector",

$$I_n \propto \left(\frac{q_c}{Q}\right)^{2(p+1-2\mu)} .$$ (69)

As long as $Q \gg k_B T/\hbar v_F$, the temperature does not affect this result. However, for higher temperatures, the interaction-dominated power law is reflected in the dependence of the peak strength on temperature rather than in its dependence on the "renormalized photon wave vector",

$$I_n \propto \left(\frac{\hbar q_c v_F}{k_B T}\right)^{2(p+1-2\mu)} .$$ (70)

7 Interband Raman Spectra Approaching Resonance

In this section, we consider the contribution of the interband modes towards the resonant Raman spectra when two subbands are occupied. We predict a new "selection rule" for the interband "SPE". It says that only the interband spin density excitations with positive group velocities can contribute as sharp, peak-like features towards the resonant Raman spectra in parallel configuration.

If the energy of the incident photons is sufficiently high, we are far from resonance and can again consider only the lowest order term in the expansion of D^{-1}. In this limit, we obtain from (43) with $i \neq j$ and choosing the zero of energy such that $\varepsilon_1 = 0$

$$N_0(q,t) = \sum_{i \neq j, \lambda} \frac{\gamma_\parallel (e_I \cdot e_O)\, \rho_{ij}^{(\lambda)}(q,t) + i\gamma_\perp |e_I \times e_O|\, \sigma_{ij}^{(\lambda)}(q,t)}{E_G + \varepsilon_i - \hbar\omega_I + \lambda\hbar v_F q/2}$$ (71)

which implies again the "classical selection rule".

The behavior of the peaks in the frequency-wave number plane to be expected for the non-resonant Raman spectra, $|E_G - \hbar\omega_I| \gg \varepsilon_i + \lambda\hbar v_F q/2$ where the denominator is independent of q has been discussed earlier [32]. Remarkably, the anti-symmetric charge excitation denoted as CDE_1^- turns out to have a very small intensity such that it cannot be observed far from resonance [29,32]. Similarly, the symmetric spin excitation SDE_1^+ can at best be observed at very small wave numbers while SDE_1^- will be present at larger wave numbers. This behavior is solely related with the q-dependence of the Bosonic modes. On the other hand, when the photon energies approach resonance, we have to take into account additionally the wave-number dependence of the energy denominator.

Although (71) is only the lowest-order term in the expansion, we nevertheless have to take into account its contribution towards the spectra when we consider photon energies approaching resonance. Indeed, for $|E_G + \varepsilon_i - \hbar\omega_I| \approx \hbar v_F q/2$ the term can contribute considerably. The behaviors in frequency-wave number space of the Raman peaks according to the zero-order term are shown in

Fig. 3a [32]. In parallel polarization, only charge density excitations are observable: CDE_1^- is expected to dominate at small q, and CDE_1^+ becomes strong for larger q. In perpendicular configuration, only SDE_1^- is expected to contribute.

Using the Bosonic representation of the Fermionic operators in (48), the first-order contribution to $N(q,t)$ is

$$N_1(q,t) = -\frac{\pi}{L}\frac{\hbar^2}{m} \sum_{i \neq j\lambda} \frac{k_F + \lambda q/2}{[E_G + \varepsilon_i - \hbar\omega_I + \lambda\hbar v_F q/2]^2} \times$$

$$\times \sum_{k,l} \left\{ \gamma_{\parallel}(e_I \cdot e_O)\left[\rho_{ll}^{(\lambda)}(q-k,t)\rho_{ij}^{(\lambda)}(k,t) + \sigma_{ll}^{(\lambda)}(q-k,t)\sigma_{ij}^{(\lambda)}(k,t) \right] + \right.$$

$$\left. + i\gamma_{\perp}\,|e_I \times e_O|\left[\rho_{ll}^{(\lambda)}(q-k,t)\sigma_{ij}^{(\lambda)}(k,t) + \sigma_{ll}^{(\lambda)}(q-k,t)\rho_{ij}^{(\lambda)}(k,t) \right] \right\}. \quad (72)$$

The polarized terms in Eq. (72) ($\propto e_I \cdot e_O$) are bi-linear in the spin density operators and yield the dominant contribution to the Raman cross section, similarly to what has been observed in the previous section for the intra-subband SDE. The important point is that they couple directly inter- and intra-subband modes in the present case.

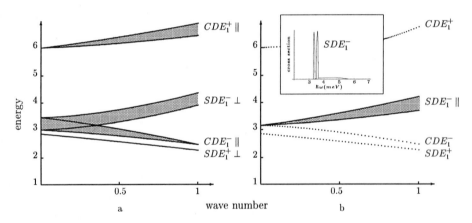

Fig. 3. (a) Behavior in frequency-wave number space of the interband resonant Raman peaks according to the zero-order term in Eq. (71) for realistic values of the model parameters; energy in meV; wave number in $10^5 \mathrm{cm}^{-1}$. (b) interband resonant Raman peaks for polarized configuration from the first-order term, Eq. (72); widths of shaded areas: intensities of the peaks in the Raman cross section. Dotted branches: intensity of Raman peaks unobservably small. Inset: Raman cross section (arbitrary units) for the interband SDE for $T = 0$, with $\hbar v_{F1} = 10^{-5}$ meV·cm, $v_{F2} = 0.6\, v_{F1}$, evaluated at $q = 0.8\ 10^5$ cm^{-1}. Notice the pronounced double peak structure which is due to the interband mode with positive group velocity (SDE_1^-). The flat background-structure between 3.5 meV and 5.5 meV is related to SDE_1^+.

By writing them in terms of the eigenmodes of the Hamiltonian, for which the time evolutions are known, and by evaluating the response function Eq. (38)

in analogy with the intraband case, we obtain a closed expression for $d^2\sigma/d\Omega d\omega$. Figure 3b shows the behavior in frequency-wave number space of the interband peaks according to the first-order term (72). The only mode which contributes in parallel polarization as a "SPE" is the anti-symmetric spin mode SDE_1^-! Due to the mixing of this interband mode which has a *positive* group velocity with the two intraband SDE with two different Fermi velocities, a double peak appears in the spectrum at zero temperature (Fig. 3 inset). The width of the peaks is roughly proportional to the difference between the group velocities of the coupled modes and they are centered on two lines with slopes equal to the two Fermi velocities v_{F1} and v_{F2}, starting near $\hbar\omega_{\sigma1}^-$. For non-zero temperatures, the two peaks will be smeared and appear as only one structure. The presence of disorder and exchange interaction will additionally broaden the peaks. The resulting SDE-structure in the parallel configuration appears roughly at the frequency of the SDE_1^-, and its intensity is predicted to increase with increasing wave number as indicated by the increasing width of the shaded area in Fig. 3b.

On the other hand, by coupling the interband spin density mode with *negative* group velocity to the two intraband spin density excitations we obtain a very flat and broad structure (Fig. 3b inset) that will not be observable in experiment.

In the terms in Eq. (72), which correspond to perpendicular polarization ($\propto e_I \times e_O$), charge and spin density operators are mixed. Again, these contribute towards the Raman cross section only with broad structures within frequency windows that are given by the differences of the group velocities of the modes that are mixed. They are not expected to be experimentally detectable. Thus, differently from the SDE, the CDE are expected to appear always only in the parallel polarization.

8   Comparison with Experiments

Comparing with experiment, we note first of all that our above Bosonization theory explains straightforwardly the spectral features of the intra- and interband excitations observed in non-resonant Raman scattering. In particular, the different dispersive behavior of the intra-subband CDE and SDE is recovered which is signature of the so-called spin-charge separation predicted by the Tomonaga-Luttinger model for the interacting one dimensional electron gas. Quantitative comparison of the calculated intra-subband CDE dispersion with experiment leads to the conclusion that the electrons in quantum wires interact via a *Coulomb potential* which is hardly screened [30].

Most importantly, a physical interpretation of the "single particle excitations" in quantum wires that appear in resonant Raman scattering has been given. Their existence has been contradicting the non-Fermi liquid nature of the one dimensional electron system and has been a disturbing puzzle for more than one decade [29]. In addition, the new "selection rule" predicted in the preceding section for the interband "SPE"-features explains why so far in experiments on quantum wires involving several subbands only "SPE" with positive group velocities have been found [33].

In order to judge, whether or not one can expect the interaction-dominated power laws of the peak intensities, predicted above, to be observable in experiment, let us consider characteristic values of parameters for samples used in recent measurements [27]. From the behavior of the intraband CDE detected by Raman scattering at small q one obtains as a rough, first estimate $g \approx 0.3$. The energy gap in the AlGaAs/GaAs hetero-structure containing the 2D electron gas is estimated to be approximately 1503 meV. The energy distance between successive subbands is approximately 3 meV. Since in the experiments reported in [27] about nine subbands are occupied, the Fermi energy is roughly between 24 and 27 meV, measured from the bottom of the lowest subband. This is confirmed by calculating the Fermi velocity from the slope of the dispersion of the SDE, and using the effective mass of the conduction band in GaAs, $m \approx 0.07m_0$. Due to the one dimensional confinement, the above 2D energy gap will be increased by a confinement energy of about 1.5 meV. The effective energy gap is then $E_G \approx 1530$ meV. The spectra in [27] are measured near $q \approx 0.8 \cdot 10^5$ cm^{-1} at a temperature of 12 K. We note that the largest photon energy for which the intraband "SPE" is observable is roughly $\hbar\omega_{\mathrm{I}}^{\max} \approx 1565$ meV, the smallest, where the peak is approximately five to ten times higher, $\hbar\omega_{\mathrm{I}}^{\min} \approx 1555$ meV. The values of Q which correspond to this range are then around $0.5\,k_{\mathrm{F}} \gg k_{\mathrm{B}}T/\hbar v_{\mathrm{F}} \approx 2 \cdot 10^{-2} k_{\mathrm{F}}$. Putting these data into (68), one finds that they are more or less consistent.

Thus, using the Tomonaga-Luttinger model, we have found strong violations of the "classical selection rules" for Raman scattering near resonance not only for the intra- but also for the interband transitions that are completely consistent with presently available experiments. Due to higher order terms which become important when the energy of the incident photons approaches the energy gap between conduction and valence band of the semiconductor, SDE modes appear in *perpendicular* as well as in *parallel* polarization of incident and scattered photons. For the latter, we have not only established their physical origin but also predicted a novel "selection rule": only interband SDE which can combine with intraband modes with similar group velocities can occur in parallel polarization as distinct peaks. On the other hand, in perpendicular configuration, no inter- and intra-subband CDE modes become observable.

Our model allows also to understand the physical reason for this selection rule: coupling between an interband SDE with positive group velocity and intraband modes which have group velocities ($\approx$ Fermi velocity) leads to a long-range correlation in time between the corresponding collective excitations. As a consequence, $\chi(q,\omega)$, the Fourier transformed correlation function, will show a pronounced structure in (q,ω)-space. On the other hand, due to opposite signs of the group velocities of the interband mode $\mathrm{SDE}_{\mathrm{I}}^{+}$ and of the two intraband excitations, the correlation in time is short-range. Its Fourier transform will be long-range in frequency. A similar qualitative argument explains also the *absence* of pronounced, sharp peaks in perpendicular polarization due to the combined spin-charge excitations which have necessarily very different group velocities. As the origin of structure in perpendicular polarization is related to spin-orbit coupling, no higher-order CDE alone can occur. Only higher-order CDE accompanied by at least one SDE are present. This physical argument suggests that the new selection rule is very probably also valid for more than two subbands.

9   Conclusions

In summary, we have developed a theory for the Raman spectra of the interacting electrons in quantum wires, which makes explicit use of the Bosonization method developed for the one dimensional Tomonaga-Luttinger model. We have shown that presently available Raman data can in principle be completely understood in terms of collective excitations which reflect the interaction. Formerly so-called "SPE" have been found to be understandable in terms of higher order SDE that appear in the polarized Raman spectra due to relaxation of the classical selection rule near resonance.

Spin charge separation, a characteristic of non-Fermi liquid behavior has been clearly identified in the Raman spectra. In addition, our theory predicts another characteristic of non-Fermi liquid behavior, namely a specific power law behavior of the intraband SDE-peak strength which should become dominant close to resonance, when $\hbar\omega_I \approx E_G$, and which would not appear, if the system was a Fermi liquid.

First quantitative estimates of parameters based on data that are presently available are consistent with our results, and suggest that Raman spectroscopy provides a quantitative tool for detecting non-Fermi liquid behavior. However, for a more precise confirmation of this conclusion, more experimental data are necessary, not only concerning the dependence on the energy of the incoming photons, but also on the temperature. Especially, data are required for quantum wires with only two occupied subbands.

In addition, the behavior of the peak strengths have to be evaluated more accurately by calculating the multiple integrals for the peak strength more precisely with numerical methods. The theory has to be generalized to quasi-one dimensional systems with more than two subbands occupied, which can also be done with numerical methods. We claim that using the present technique, by comparing theory with experiment, a very detailed picture of the electron-electron interaction in quantum wires, and the corresponding collective excitations, can be obtained.

In conclusion, previous and present experimental and theoretical findings indicate that the observed features in the Raman spectra of quantum wires are completely understandable in terms of the collective intra- and interband spin and charge excitations characteristic of coupled Tomonaga-Luttinger liquids. This represent in our opinion strong evidence that the Tomonaga-Luttinger model in connection with Bosonization correctly describes the collective modes in quantum wires for $q \neq 0$ which are dominated by the Coulomb interaction.

It remains a puzzle and a challenge to experiment why such strong interaction effects have been found notoriously difficult to be detected in transport measurements on quantum wires.

Acknowledgements: We acklowledge with pleasure several useful discussions concerning experiments with Detlef Heitmann, Alessandro Goñi, Klaus von Klitzing, Bernard Jusserand and Aron Pinzcuk. It is also a great pleasure to thank Boris Altshuler, Franco Napoli and Eros Mariani for continuous constructive contributions. The work has been financially supported by the Ministero dell'Università Ricerca Scientifica e Tecnologica via Cofinanziamento 98, by the Deutsche Forschungsgemeinschaft via SFB 508 "Quantenmaterialien" of the Universität Hamburg and has been carried out within the European Network TMR (FMRX-CT96-0042 and FMRX-CT98-0180).

References

1. S. Tomonaga, Prog. Theor. Phys. **5**, 544 (1950).
2. J. M. Luttinger, J. Math. Phys. **4**, 1154 (1963).
3. F. D. M. Haldane, J. Phys. C **14**, 2585 (1981).
4. J. Sólyom. Adv. Phys. **28**, 201 (1979).
5. J. Voit, Rep. Progr. Phys. **57**, 977 (1995).
6. M. Sassetti, in: *Quantum Transport in Semiconductor Submicron Structures*, ed. by B. Kramer, NATO ASI Ser. E **326** (Kluwer, Dordrecht 1996), p. 95.
7. H. J. Schulz, Phys. Rev. Lett. **71**, 1864 (1993); H. J. Schulz, in *Mesoscopic Quantum Physics*, edited by E. Akkermans, G. Montambaux, J. L. Pichard, and J. Zinn-Justin (Elsevier, New York 1995), p. 533.
8. T. Ogawa, Physica B **249-251**, 185 (1998).
9. D. G. Clarke, S. P. Strong, P. W. Anderson, Phys. Rev. Lett. **72**, 3218 (1994).
10. S. Tarucha, T. Honda and T. Saku, Sol. St. Commun. **94**, 413 (1995).
11. A. Yacoby, H. L. Störmer, N. S. Wingreen, L. N. Pfeiffer, K. W. Baldwin and K. W. West, Phys. Rev. Lett. **77**, 4612 (1996).
12. M. Bockrath, D. H. Cobden, J. Lu, A. G. Rinzler, R. E. Smolley, L. Balents, P. L. McEuen, Nature **397**, 598 (1999).
13. X. G. Wen, Phys. Rev. **B43**, 11025 (1991); Int. J. Mod. Phys. **B6**, 1711 (1992).
14. C. de Chamon, X. G. Wen, Phys. Rev. Lett. **70**, 2605 (1993).
15. K. Moon, H. Yi, C. L. Kane, S. M. Girvin, M. P. A. Fisher, Phys. Rev. Lett. **71**, 4381 (1993).
16. F. P. Millikan, C. P. Umbach and R. A. Webb, Sol. St. Commun. **97**, 309 (1996).
17. A. M. Chang, L. N. Pfeiffer, and K. W. West, Phys. Rev. Lett. **77**, 2538 (1996).
18. M. Grayson, D. C. Tsui, L. N. Pfeiffer, K. W. West, A. M. Chang, Phys. Rev. Lett. **80**, 1062 (1998).
19. G. M. Auslaender, A. Yacoby, R. de Picciotto, K. W. Baldwin, L. N. Pfeiffer and K. W. West, cond-mat/9909138.
20. D. C. Hamilton, A.L. McWhorter, in: *Light Scattering Spectra of Solids*, edited by G. B. Wright, p. 309 (Springer Verlag, Berlin 1969).
21. F. A. Blum, Phys. Rev. **B1**, 1125 (1970).
22. M. V. Klein, in: *Light Scattering in Solids*, edited by M. Cardona, Topics in Appl. Phys. **8**, p. 147 (Springer Verlag, Berlin 1975).
23. A. R. Goñi, A. Pinczuk, J. S. Weiner, J. M. Calleja, B. S. Dennis, L. N. Pfeiffer and K. W. West, Phys. Rev. Lett. **67**, 3298 (1991).
24. A. R. Goñi et al., in *Phonons in Semiconductor Nanostructures*, edited by J. P. Leburton, J.Pascual and C. S. Torres, (Plenum, New York 1993), p. 287.

25. A. Schmeller, A. R. Goñi, A. Pinczuk, J. S. Weiner, J. M. Calleja, B. S. Dennis, L. N. Pfeiffer and K. W. West, Phys. Rev. B49, 14778 (1994).

26. R. Strenz, U. Bockelmann, F. Hirler, G. Abstreiter, G. Böhm and G. Weimann, Phys. Rev. Lett. 73, 3022 (1994).

27. C. Schüller, G. Biese, K. Keller, C. Steinebach, D. Heitmann, P. Grambow and K. Eberl, Phys. Rev. B54, R17304 (1996).

28. F. Perez, B. Jusserand, and B. Etienne, Physica E, in press (1999).

29. M. Sassetti, B. Kramer, Phys. Rev. Lett. 80, 1485 (1998); Eur. Phys. J. B 4, 357 (1998).

30. M. Sassetti, B. Kramer, D. Fichtner, C. Schüller, E. Ulrichs, C. Steinebach, D. Heitmann, in: *The Physics of Semiconductors*, edited by D. Gershoni (World Scientific Publishers, Singapore 1999), CD-version, VII B 19.

31. G. Cuniberti, M. Sassetti and B. Kramer, Phys. Rev. B57, 1515 (1998)

32. M. Sassetti, F. Napoli, and B. Kramer, Phys. Rev. B 59, 7297 (1999); Eur. Phys. J. B, in press (1999).

33. E. Mariani, M. Sassetti and B. Kramer, Ann. Phys. (Leipzig) 8, 161 (1999).

34. T. Dittrich, B. Kramer, G. L. Ingold, P. Hänggi, G. Schön and W. Zwerger, *Quantum Transport and Dissipation.* (Wiley-VCH, Weinheim 1997).

Part II

Transport and Interactions
in Zero and Two Dimensions

An Introduction to Real-Time Renormalization Group

Herbert Schoeller[1,2]

[1] Forschungszentrum Karlsruhe, Institut für Nanotechnologie, 76021 Karlsruhe, Germany
[2] Institut für Theoretische Festkörperphysik, Universität Karlsruhe, 76128 Karlsruhe, Germany

1 Introduction

This article presents a tutorial introduction to a recently developed real-time renormalization group method [1]. It describes nonequilibrium properties of discrete quantum systems coupled linearly to an environment. We illustrate the technique by a simple and exactly solvable model: A quantum dot consisting of a single non-degenerate level coupled to two reservoirs. The article is intended for advanced students. Besides elementary quantum mechanics and statistical mechanics, it requires knowledge of second quantization and Wick's theorem. The latter topics can be learned easily from standard textbooks, see e.g. Ref. [2].

Renormalization group (RG) methods are standard tools to describe various aspects of condensed matter problems beyond perturbation theory [3]. Many impurity problems have been treated by numerical RG with excellent results both for thermodynamic quantities and spectral densities [4,5]. These RG techniques, however, cannot describe nonequilibrium properties like the nonlinear conductance, the nonequilibrium stationary state, or the full time development of an initially out-of-equilibrium state. To address these aspects we present here a perturbative RG method, formulated for strongly correlated quantum systems with a finite number of states coupled linearly to external heat or particle reservoirs. Examples are: spin boson models, molecules interacting with electrodynamic fields, generalized Anderson-impurity models, quantum dot devices, magnetic nanoparticles interacting with phonons, etc.. Fundamentally new, we generate non-Hamiltonian dynamics during RG, which captures the physics of finite life times and dissipation. Furthermore, no initial or final cutoff in energy or time space is needed, i.e., large and small energy scales are accounted for correctly like in flow-equation methods [6]. Although correlation functions can also be studied, physical quantities like spin and charge susceptibilities or the current can be calculated directly without the need of nonequilibrium Green's functions.

The purpose of our RG technique is to describe quantum fluctuations which are induced by strong coupling between a small quantum system and an environment. There are several recent experiments which show the importance of quantum fluctuations in metallic single-electron transistors [7] and semiconductor quantum dots [8] (see [9] for an overview over theoretical papers). Due to the renormalization of resistance and local energy excitations, anomalous line shapes of the conductance have been observed, which can not be explained by golden-rule theories. For applications of the real-time RG to these cases we refer to Rfs. [1,10].

Here we want to illustrate the method by an exactly solvable model, namely a quantum dot with one non-degenerate state with energy ϵ coupled to two reservoirs ($r = L, R$). The Hamiltonian $H = H_R + H_0 + H_T$ consists of three parts, corresponding to the reservoirs, the dot, and tunneling

$$H_R = \sum_r H_r = \sum_{kr} \epsilon_{kr} a_{kr}^\dagger a_{kr} , \qquad (1)$$

$$H_0 = \epsilon \, c^\dagger c , \qquad (2)$$

$$H_T = \sum_{kr} \left\{ T_r a_{kr}^\dagger c + T_r^* c^\dagger a_{kr} \right\} . \qquad (3)$$

Nonequilibrium is taken into account by describing the electrons in the reservoirs by Fermi distribution functions with different electrochemical potentials μ_r. Tunneling is switched on suddenly at the initial time t_0, i.e. initially the density matrix $\rho(t_0) = \rho_0$ decouples into an equilibrium part for each reservoir, $\rho_r = Z_r^{-1} e^{-\beta(H_r - \mu_r N_r)}$, and an arbitrary initial distribution $p(t_0) = p_0$ for the dot

$$\rho_0 = p_0 \rho_{res} = p_0 \rho_L \rho_R . \qquad (4)$$

The aim is to calculate the time evolution of the reduced density matrix of the dot, $p(t) = Tr_{res} \rho(t)$, and the tunneling current $\langle I_r \rangle(t)$ flowing from reservoir r to the dot. The tunneling current operator is given by [1]

$$I_r = -e\dot{N}_r = ie \sum_k \left\{ T_r a_{kr}^\dagger c - T_r^* c^\dagger a_{kr} \right\} . \qquad (5)$$

The solution of the above quadratic Hamiltonian is trivial since all degrees of freedoms can easily be integrated out. Doing this by using Wick's theorem for all field operators within the Keldysh formalism, one can easily solve the full nonequilibrium problem [11]. However, except for having solved a special and almost trivial problem, we would not have gained anything for solving more general problems of dissipative quantum mechanics. Usually, the local system can not be integrated out due to interaction terms or spin degrees of freedom. Therefore, we try to proceed differently. We will only integrate out the reservoirs and keep the dot degrees of freedom explicitly. This is always possible for an effectively noninteracting bath. As a result, we get an effective theory in terms of the local degrees of freedom, expressed by a formally exact kinetic equation for the reduced density matrix of the dot. For the special Hamiltonian (1)-(3), we solve this equation exactly. Furthermore, we will also develop a renormalization group method to solve the kinetic equation. We show that the RG equations describe the same exact solution. The important point is that both steps, i.e.

---

[1] Throughout this work we set $\hbar = k_B = 1$ and use $e < 0$

(a) setting up the kinetic equation, and (b) setting up the RG equations, are not specific to the above Hamiltonian but can be applied to any discrete quantum system coupled linearly to an environment. The only difference is that, for most problems, the resulting RG equations have to be solved numerically. In conclusion, the above Hamiltonian serves as a test example to illustrate the RG technique and to demonstrate that it is well-defined and useful.

2   Diagrammatic Language

2.1   Diagrams on the Keldysh Contour

We start by introducing some convenient notations. The index $\mu = \eta r$ labels the possible tunneling processes between reservoirs and dot. $\eta = \pm$ indicates tunneling in/out, and $r = L, R$ specifies the reservoir. We define the following reservoir and dot operators

$$j_{-r} = \sum_k T_r a_{kr}^\dagger \,, \quad j_{+r} = \sum_k T_r^* a_{kr} \,, \tag{6}$$

$$g_{-r} = c \,, \quad g_{+r} = c^\dagger \,. \tag{7}$$

We denote the two possible states of the dot by $s = 0, 1$ with energies $E_0 = 0$ and $E_1 = \epsilon$. The Hamiltonian (1)-(3) becomes

$$H_R = \sum_r \epsilon_{kr} a_{kr}^\dagger a_{kr} \,, \tag{8}$$

$$H_0 = \sum_s E_s |s\rangle\langle s| \,, \tag{9}$$

$$H_T = \sum_\mu : g_\mu j_\mu := \sum_r \{g_{+r} j_{+r} + j_{-r} g_{-r}\} \,, \tag{10}$$

where the symbol $: \cdots :$ denotes normal ordering of Fermi field operators but without sign change when two operators are interchanged. The current operator (5) for the left reservoir is given by

$$I_L = \sum_\mu : i_\mu j_\mu : \,, \quad i_\mu = -ie\eta g_\mu \delta_{rL} \,, \quad (\mu = \eta r) \,, \tag{11}$$

and a corresponding equation for I_R.

The time evolution of an arbitrary observable a follows from the von Neumann equation

$$\langle a \rangle(t) = Tr\, a\rho(t) = Tr\, a\, e^{-iH(t-t_0)} \rho_0 e^{iH(t-t_0)} \tag{12}$$

$$= Tr e^{iH(t-t_0)} a\, e^{-iH(t-t_0)} \rho_0 \,, \tag{13}$$

where, in the last step, we have used cyclic invariance under the trace. To get a matrix element of the reduced density matrix of the dot, $p(t)_{ss'} = \langle a \rangle(t)$, we need $a = |s'\rangle\langle s|$. For the current, we take $a = I_L = \sum_\mu : i_\mu j_\mu :$.

To integrate out the reservoirs, we expand the propagators in tunneling and apply Wick's theorem to the reservoir degrees of freedom. We introduce the interaction picture

$$b(t) = e^{i(H_R+H_0)(t-t_0)} \, b \, e^{-i(H_R+H_0)(t-t_0)} \,, \tag{14}$$

and obtain

$$e^{iH(t-t_0)} \, a \, e^{-iH(t-t_0)} = \tilde{T} \, e^{i\int_{t_0}^t dt' \, H_T(t')} \, a(t) \, T \, e^{-i\int_{t_0}^t dt' \, H_T(t')} \,, \tag{15}$$

where T and $\tilde{T}$ denote the time-ordering and anti-time-ordering operators, respectively. Inserting (15) in (13) and expanding in H_T gives a series of terms which we visualize diagrammatically, see Fig. 1. The upper (lower) line corresponds to the forward (backward) propagator. The diagram shown corresponds to the following expression

$$i^2(-i)^2 \, Tr \, H_T(t_3) H_T(t_2) \, a(t) \, H_T(t_1) H_T(t_4) \, \rho_0 \,. \tag{16}$$

We see that the operators are ordered along a closed time path (Keldysh contour), as shown in Fig. 1.

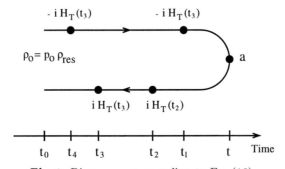

Fig. 1. Diagram corresponding to Eq. (16).

The next step is to insert $H_T = \sum_\mu : g_\mu j_\mu :$ for the tunneling Hamiltonian and $\rho_0 = \rho_0 \rho_{res}$ for the initial density matrix. We use the short-hand notation $g_i = g_{\mu_i}(t_i)$ and $j_i = j_{\mu_i}(t_i)$, and decompose each diagram into a dot and reservoir part. If $a = |s'><s|$ is a dot operator, we get from (16)

$$i^2(-i)^2 \sum_{\mu_1\mu_2\mu_3\mu_4} \{Tr_0 \, g_3 g_2 \, a(t) \, g_1 g_4 \, \rho_0\} \, \{Tr_{res} \, j_3 j_2 j_1 j_4 \, \rho_{res}\} \,, \tag{17}$$

whereas, if $a = I_L = \sum_\mu : i_\mu j_\mu :$ is the current operator, we get

$$i^2(-i)^2 \sum_{\mu\mu_1\mu_2\mu_3\mu_4} \{Tr_0 \, g_3 g_2 \, i_\mu(t) \, g_1 g_4 \, \rho_0\} \, \{Tr_{res} \, j_3 j_2 \, j_\mu(t) \, j_1 j_4 \, \rho_{res}\} \,. \tag{18}$$

Here, Tr_0 (Tr_{res}) denotes the trace over the dot (reservoir) degrees of freedom. The reader can convince himself very easily that this factorization does not imply any additional minus signs from commutation of Fermi operators. The reason is the quadratic form of the tunneling Hamiltonian.

The trace over the reservoirs can be calculated easily by using Wick's theorem [2]. As a result, we can decompose any average over products of reservoir field operators into a sum over products of pair contractions. Denoting by $\langle \ldots \rangle$ the average over the reservoirs, we get for the reservoir part of (17)

$$\langle j_3 j_2 j_1 j_4 \rangle = \overset{\sqcap\;\sqcap}{j_3 j_2 j_1 j_4} + \overset{\sqcap\!\sqcap}{j_3 j_2 j_1 j_4} + \overset{\sqcap\;\;\sqcap}{j_3 j_2 j_1 j_4}$$

$$= \langle j_3 j_2 \rangle \langle j_1 j_4 \rangle - \langle j_3 j_1 \rangle \langle j_2 j_4 \rangle + \langle j_3 j_4 \rangle \langle j_2 j_1 \rangle . \qquad (19)$$

Each pair contraction corresponds to an equilibrium average over two reservoir field operators. If two contractions intersect proper minus signs have to be taken into account due to Fermi statistics. A pair contraction $\langle j_\mu(t) j_{\mu'}(t') \rangle$ depends on the relative time $t - t'$ and is only non-zero for $\mu' = \bar{\mu}$, with $\bar{\mu} = -\eta r$ (if $\mu = \eta r$). We define

$$\gamma_\mu(t) = \langle j_{\bar{\mu}}(t) j_\mu \rangle , \qquad (20)$$

and get, by using the definition (6)

$$\gamma_\mu(t) = \sum_k |T_r|^2 \begin{cases} \langle a_{kr}^\dagger(t) a_{kr} \rangle & \text{for} \quad \eta = + \\ \langle a_{kr}(t) a_{kr}^\dagger \rangle & \text{for} \quad \eta = - \end{cases}$$

$$= \frac{1}{2\pi} \int dE \, \Gamma_r(E) e^{i\eta Et} f^\eta (E - \mu_r) \qquad (21)$$

$$\cong \frac{\Gamma_r}{2\pi} \int dE \, e^{\eta(E-\mu_r)/D} e^{i\eta Et} f_r^\eta(E) . \qquad (22)$$

Here we have defined

$$\Gamma_r \cong \Gamma_r(E) = 2\pi \sum_k |T_r|^2 \delta(E - \epsilon_{kr}) = 2\pi |T_r|^2 N_r(E) , \qquad (23)$$

with $N_r(E)$ being the density of states of reservoir r. We define $f^+ = f$, $f^- = 1 - f$, and $f_r^\eta(E) = f^\eta(E - \mu_r)$, with $f(E) = 1/(\exp(\beta E) + 1)$ being the Fermi function, and $\beta = 1/T$ the inverse temperature. We take the density of states $N_r(E)$ independent of energy and regularize the integral (21) by introducing an exponential convergence factor $e^{\eta(E-\mu_r)/D}$. Here, D corresponds to a high energy cutoff. In the end, we will send $D \to \infty$. Performing the integral (22) gives

$$\gamma_\mu(t) = \frac{-i\Gamma_r e^{i\eta\mu_r t}}{2\beta \sinh[\pi(t - i/D)/\beta]} . \qquad (24)$$

Furthermore, we define

$$\gamma_r^\eta(t) = \gamma_\mu(\eta t) , \qquad (25)$$

and note the important property

$$\lim_{D \to \infty} \left\{ \gamma_r^+(t) + \gamma_r^-(t) \right\} = \Gamma_r \delta(t) \;, \tag{26}$$

which follows directly from (22).

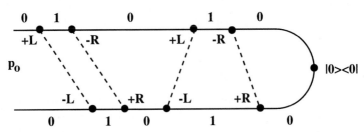

Fig. 2. Diagram after the application of Wick's theorem. On the propagators the states $s = 0, 1$ of the dot are shown. At the vertices we have indicated the index $\mu = \eta r$.

We indicate the pair contractions diagrammatically by connecting the corresponding vertices by a dashed line, see Fig. 2. The remaining part for the dot degrees of freedom still remains, see Eq. (17). We calculate this part by inserting intermediate states of the dot between the operators. These states are indicated in Fig. 2 between the tunneling vertices. The reservoirs are already integrated out, so the tunneling vertices correspond to the dot operators g_{μ_i} (with an additional factor $\mp i$ for a vertex on the upper (lower) propagator). Between the tunneling vertices we have the free time evolution of the dot, i.e. for a propagation of state s from t_2 to t_1, we get an exponential factor $e^{-iE_s(t_1 - t_2)}$.

Our diagrammatic language provides an effective description in terms of the dot degrees of freedom. The presence of the reservoirs is reflected by the coupling of the tunneling vertices by the free Green's function of the reservoirs. In particular, this means that the forward and backward propagator are no longer independent but are coupled by reservoir lines. We will see in section 3.1 that this leads to rates in a kinetic equation for $p(t)$.

2.2   Superoperator Notation

In this section we will replace the double-propagator diagrams on the Keldysh contour by a convenient matrix notation. This provides a very compact and analytic way to express diagrams by formulas.

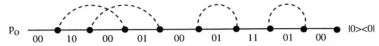

Fig. 3. Diagram in the double-state representation. It results from Fig. 2 by taking the upper and lower line together to one single line.

Instead of considering two propagators and specifying the dot states on each propagator separately, we can formally take both propagators together to one line and specify the state on this new propagator by a double state (s, s'), see Fig. 3. Here, the first (second) state corresponds to the state on the upper (lower) propagator. By this trick we have lost the information wether a tunneling vertex lies on the upper or lower propagator. To recover this we add to the index μ of the tunneling vertex operator an additional index $p = \pm$, which indicates wether g_μ acts on the upper (lower) propagator. The new vertex is denoted by G_μ^p with matrix elements

$$(G_\mu^+)_{s_1 s_1', s_2 s_2'} = -i(g_\mu)_{s_1 s_2} \delta_{s_1' s_2'} \tag{27}$$

$$(G_\mu^-)_{s_1 s_1', s_2 s_2'} = i\delta_{s_1 s_2} (g_\mu)_{s_2' s_1'} . \tag{28}$$

We have included the factors $\mp i$ for a vertex on the upper (lower) propagator into the definition of G_μ^p. The free time evolution between the vertices is given by a factor $e^{-i(E_s - E_{s'})(t_1 - t_2)}$, where (s, s') indicates the double state on the line. This can be written in operator form as $(e^{-iL_0(t_1 - t_2)})_{ss', ss'}$, with

$$(L_0)_{s_1 s_1', s_2 s_2'} = \delta_{s_1 s_2} \delta_{s_1' s_2'} (E_{s_1} - E_{s_1'}) . \tag{29}$$

Finally, a contraction connecting a vertex $G_{\mu'}^{p'}$ at time t' with a vertex G_μ^p at time $t > t'$, is denoted by $\gamma_{\mu\mu'}^{pp'}(t - t')$. Using the definition (20) and the fact that operators on the lower propagator act always later than those on the upper propagator, we obtain

$$\gamma_{\mu\mu'}^{pp'}(t) = \delta_{\bar\mu, \mu'} \begin{cases} \langle j_\mu(t) j_{\bar\mu} \rangle = \gamma_{\bar\mu}(t) & \text{for} \quad p' = + \\ \langle j_{\bar\mu} j_\mu(t) \rangle = \gamma_\mu(-t) & \text{for} \quad p' = - \end{cases} \tag{30}$$

$$= \delta_{\bar\mu, \mu'} \frac{-i\Gamma_r e^{-i\eta\mu_r t}}{2\beta \sinh[\pi(p't - i/D)/\beta]} , \tag{31}$$

where we used the result (24) in the second step.

The double-state matrices L_0 and G_μ^p are called superoperators in the sense that they act on single-state matrices, i.e. on ordinary operators. If b is an ordinary operator, we can define L_0 and G_μ^p by

$$L_0 b = [H_0, b] \quad , \quad G_\mu^+ b = -i g_\mu b \quad , \quad G_\mu^- b = i b g_\mu . \tag{32}$$

Within the superoperator notation and using $a = |s'\rangle\langle s|$, an arbitrary diagram for $p_{ss'}(t)$ can be written as

$$p_{ss'}(t) \quad \rightarrow \quad \left\{ e^{-iL_0(t-t_0)} \overbrace{G_1 G_2 \ldots \ldots G_n}\, p_0 \right\}_{ss'} . \tag{33}$$

The time dependence of $G_i = G_{\mu_i}^{p_i}(t_i)$ is defined by

$$G_\mu^p(t) = e^{iL_0(t-t_0)} G_\mu^p e^{-iL_0(t-t_0)} . \tag{34}$$

All operators $G_1 \ldots G_n$ are coupled in all possible ways by reservoir pair contractions, as indicated in (33). Implicitly we assume summation over $\mu_1 \ldots \mu_n$ and $p_1 \ldots p_n$, together with the integration over the time variables $t_1 \ldots t_n$ with $t > t_1 > t_2 > \cdots > t_n > t_0$.

If $a = I_L = \sum_\mu : i_\mu j_\mu :$ corresponds to the current operator, we get

$$\langle : i_\mu j_\mu : \rangle(t) \quad \rightarrow \quad Tr_0\, i_\mu \left\{ e^{-iL_0(t-t_0)}\, \overbrace{G_1 G_2}\overbrace{\ldots\ldots} G_n\, p_0 \right\} . \tag{35}$$

In comparison to (33), we need an additional pair contraction to the current vertex. In order to treat the boundary vertex i_μ as well within the superoperator notation, we define a superoperator I_μ^p by

$$I_\mu^+ b = i_\mu b/2 \quad , \quad I_\mu^- b = b\, i_\mu/2 , \tag{36}$$

with matrix elements

$$(I_\mu^+)_{s_1 s_1', s_2 s_2'} = \frac{1}{2}(i_\mu)_{s_1 s_2}\delta_{s_1' s_2'} \quad , \quad (I_\mu^-)_{s_1 s_1', s_2 s_2'} = \frac{1}{2}\delta_{s_1 s_2}(i_\mu)_{s_2' s_1'} . \tag{37}$$

Using cyclic invariance under the trace, we get for (35)

$$\boxed{\langle : i_\mu j_\mu : \rangle(t) \quad \rightarrow \quad Tr_0\, I_t\, \overbrace{G_1 G_2}\overbrace{\ldots\ldots} G_n\, p_0 , } \tag{38}$$

where $I_t = I_\mu^p(t)$, and the interaction picture is defined by

$$I_\mu^p(t) = I_\mu^p\, e^{-iL_0(t-t_0)} . \tag{39}$$

Eqs. (33) and (38) are the central result of this section. They relate the reduced density matrix of the dot and the average current to diagrammatic expressions in a very compact and analytic way. It turns out that the usage of superoperators simplifies the notation considerably. We will see in sections 3 and 4 that the derivation of kinetic equations and renormalization group equations is very transparent in this language. However, one should always keep in mind that the usage of superoperators is only a formal trick to find a convenient matrix notation. Therefore, we have set up the diagrammatic representation in terms of the Keldysh contour first, and then, in a second step, introduced the superoperators. Of course it is also possible to start directly with superoperators [1], which provides a more compact and shorter way to arrive at Eqs. (33) and (38). However, for pedagogical and physical reasons, we did not proceed in this way here. The diagrams on the Keldysh contour reveal better that there are different kinds of terms which have to be distinguished very carefully from a physical point of view. Reservoir lines connecting the upper and lower propagator correspond to rates, they change the state of the dot simultaneously on the upper *and* the lower propagator. This describes a transition from one diagonal matrix element of the reduced density matrix $p(t)$ to another one. Such processes can not be expressed on a Hamiltonian level and lead basically to the physics of dissipation. In contrast, reservoir contractions which connect vertices within the upper or lower propagator describe renormalization and broadening of levels.

3   Kinetic Equation

3.1   General Approach

In this section we will derive a self-consistent equation for the reduced density matrix $p(t)$ of the dot, together with an expression for the average current. To achieve this it is essential to distinguish in Eqs. (33) and (38) between connected and disconnected parts. From (33) we see that any diagram for $p(t)$ can be written in the form

$$e^{-iL_0(t-t_1)} \left(A_1 G \ldots G B_2\right)_{con} e^{-iL_0(t_2-t_3)} \left(A_3 G \ldots G B_4\right)_{con} \ldots$$
$$\ldots e^{-iL_0(t_{2n-2}-t_{2n-1})} \left(A_{2n-1} G \ldots G B_{2n}\right)_{con} e^{-iL_0(t_{2n}-t_0)} p_0 . \quad (40)$$

Here, $(A_i G \ldots G B_{i+1})_{con}$ denotes a sequence of vertices between t_{i+1} and t_i which are coupled by pair contractions in such a way that any vertical cut between t_{i+1} and t_i will cross some contraction. We define such a block as a connected part of a diagram. E.g. Fig. 3 shows a sequence of three connected blocks. The boundary vertices A and B are identical to G, i.e. $A_\mu^p = B_\mu^p = G_\mu^p$, but the interaction picture is defined differently

$$A_\mu^p(t) = A_\mu^p e^{-iL_0(t-t_0)} \quad , \quad B_\mu^p(t) = e^{iL_0(t-t_0)} B_\mu^p . \quad (41)$$

The reason is that we want the connected part $(A_i G \ldots G B_{i+1})_{con}$ of a diagram to depend only on the relative time argument $t_i - t_{i+1}$. Furthermore, we distinguish the boundary vertices A_μ^p and B_μ^p from G_μ^p since, within the renormalization group procedure developed in section 4, the boundary vertices renormalize differently.

We define the sum over all connected diagrams between t' and t by the kernel $\Sigma(t - t')$

$$\Sigma(t - t') \quad \rightarrow \quad (A_t G_1 G_2 \ldots G_{2n} B_{t'})_{con} . \quad (42)$$

We note the important property that the kernel $\Sigma(t - t')$ is independent of the initial time t_0 since all exponential factors $e^{\pm iL_0 t_0}$ arising from the interaction picture cancel within Σ. Thus, in order to calculate Σ, we can set $t_0 = 0$ in the definition of the interaction picture of A, B, and G, see Eqs. (34) and (41).

Using the definition (42) for Σ in (40), we obtain

$$p(t) = e^{-iL_0(t-t_0)} p_0 + \sum_{n=1}^{\infty} \int_{t_0}^{t} dt_1 \int_{t_0}^{t_1} dt_2 \cdots \int_{t_0}^{t_{2n-1}} dt_{2n}$$
$$e^{-iL_0(t-t_1)} \Sigma(t_1 - t_2) e^{-iL_0(t_2-t_3)} \Sigma(t_3 - t_4) \ldots$$
$$\ldots e^{-iL_0(t_{2n-2}-t_{2n-1})} \Sigma(t_{2n-1} - t_{2n}) e^{-iL_0(t_{2n}-t_0)} p_0 . \quad (43)$$

Differentiating with respect to time gives the kinetic equation

$$\dot{p}(t) + iL_0 p(t) = \int_{t_0}^{t} dt' \Sigma(t - t') p(t') .$$ (44)

Since the r.h.s. of this equation is a convolution in time space, we can formally solve this equation in Laplace space. We define the Laplace transform by

$$\tilde{p}(z) = \int_{t_0}^{\infty} dt\, e^{izt} p(t) \quad , \quad \tilde{\Sigma}(z) = \int_0^{\infty} dt\, e^{izt} \Sigma(t) ,$$ (45)

and get from (44) the solution

$$\tilde{p}(z) = \frac{i}{z - L_0 - i\tilde{\Sigma}(z)} p_0 .$$ (46)

The time dependence $p(t)$ follows by reversing the Laplace transform

$$p(t) = \lim_{\eta \to 0} \frac{1}{2\pi} \int_{-\infty+i\eta}^{\infty+i\eta} dz\, e^{-izt} \tilde{p}(z) = \lim_{\eta \to 0} \frac{1}{2\pi} \int_{-\infty}^{\infty} d\omega\, e^{-i\omega t} \tilde{p}(\omega + i\eta) .$$ (47)

We remark that $\tilde{p}(z)$, defined by (45), is analytic in the upper half plane since $p(t)$ will approach a stationary value for $t \to \infty$. Thus, within the integration region of (47) the integrand is well-defined. We see that the integral kernel $\tilde{\Sigma}(z)$ is the central object which has to be calculated. The full time evolution of $p(t)$ out of an arbitrary nonequilibrium state can be obtained once $\tilde{\Sigma}(z)$ is known for all $z = \omega + i0^+$. The calculation of $\tilde{\Sigma}(z)$ will be the subject of the renormalization group approach described in section 4.

The stationary state is defined by

$$p_{st} = \lim_{t \to \infty} p(t) = -i \lim_{z \to i0^+} z\, \tilde{p}(z) .$$ (48)

Multiplying (46) by $z[z - L_0 - i\tilde{\Sigma}(z)]$ and taking the limit $z \to i0^+$, we see that the stationary state is the eigenvector of $L_0 + i\tilde{\Sigma}(i0^+)$ with eigenvalue zero

$$[L_0 + \tilde{\Sigma}(i0^+)]\, p_{st} = 0 .$$ (49)

The density matrix $p(t)$ is hermitian. In Laplace space this is equivalent to

$$\tilde{p}(z)_{ss'}^{*} = \tilde{p}(-z^*)_{s's} .$$ (50)

This follows from the solution (46) by using the symmetry relations

$$(iL_0)_{s_1 s_1', s_2 s_2'}^{*} = (iL_0)_{s_1' s_1, s_2' s_2} \quad , \quad (G_\mu^p)_{s_1 s_1', s_2 s_2'}^{*} = (G_{\bar\mu}^{\bar p})_{s_1' s_1, s_2' s_2} ,$$ (51)

where $\bar{p} = -p$. They follow directly from (27), (28), (29), and the hermiticity of the Hamiltonian. The consequence for the kernel (42) is

$$\Sigma(t)^*_{s_1 s'_1, s_2 s'_2} = \Sigma(t)_{s'_1 s_1, s'_2 s_2} \quad , \quad \tilde{\Sigma}(z)^*_{s_1 s'_1, s_2 s'_2} = \tilde{\Sigma}(-z^*)_{s'_1 s_1, s'_2 s_2} \quad . \tag{52}$$

Using these relations together with the hermiticity of the initial density matrix p_0, we find directly (50) from (46).

To prove conservation of probability $Tr_0 p(t) = 1$, we first note that

$$\sum_s (L_0)_{ss,\cdot\cdot} = \sum_{sp} (A^p_\mu)_{ss,\cdot\cdot} = \sum_{sp} (B^p_\mu)_{ss,\cdot\cdot} = \sum_{sp} (G^p_\mu)_{ss,\cdot\cdot} = 0 \quad . \tag{53}$$

Applying this to (42) together with the fact that the contraction connected to the boundary operator A^p_μ does not depend on p, we find the same property for the kernel

$$\sum_s \Sigma(t)_{ss,\cdot\cdot} = \sum_s \tilde{\Sigma}(z)_{ss,\cdot\cdot} = 0 \quad . \tag{54}$$

Applying Tr_0 to the kinetic equation (44), and using the properties (53) and (54), we find $d/dt Tr_0 p(t) = 0$ which proves conservation of probability.

To calculate the current we proceed analogously. From (38) and (40) we see that any diagram for the average current can be written as

$$Tr_0 \left(I_t\, G \dots G B_{t'} \right)_{con} p(t') \quad . \tag{55}$$

The first connected block contains the current vertex. The sum over all connected diagrams containing the current vertex is denoted by $\Sigma_I(t - t')$

$$\boxed{\quad \Sigma_I(t - t') \quad \rightarrow \quad (I_t G_1 G_2 \dots G_{2n} B_{t'})_{con} \quad . \quad} \tag{56}$$

The only difference to (42) is that the boundary vertex A^p_μ has been replaced by the current vertex I^p_μ. From (55) and (56) we find

$$\boxed{\quad \langle I_L \rangle(t) = \sum_\mu \langle : i_\mu j_\mu : \rangle = \int_{t_0}^t dt'\, Tr_0\, \Sigma_I(t - t')\, p(t') \quad . \quad} \tag{57}$$

In Laplace space we get

$$\langle \tilde{I}_L \rangle(z) = Tr_0 \tilde{\Sigma}_I(z)\, \tilde{p}(z) \quad , \tag{58}$$

and the stationary solution follows from

$$\langle I_L \rangle_{st} = Tr_0\, \tilde{\Sigma}_I(i0^+)\, p_{st} \quad . \tag{59}$$

The expectation value of the current is real, i.e. $\langle \tilde{I}_L \rangle(z)^* = \langle \tilde{I}_L \rangle(-z^*)$. This can be seen from the solution (58) by using the symmetry relations

$$(I^p_\mu)^*_{s_1 s'_1, s_2 s'_2} = (I^{\bar{p}}_{\bar{\mu}})_{s'_1 s_1, s'_2 s_2} \quad , \tag{60}$$

and

$$\Sigma_I(t)^*_{s_1 s'_1, s_2 s'_2} = \Sigma_I(t)_{s'_1 s_1, s'_2 s_2} \quad , \quad \tilde{\Sigma}_I(z)^*_{s_1 s'_1, s_2 s'_2} = \tilde{\Sigma}_I(-z^*)_{s'_1 s_1, s'_2 s_2} \quad . \tag{61}$$

3.2   Exact Solution

For the model of a single non-degenerate dot state, the kinetic equation and
the current formula can be solved exactly. We derive this solution here and will
check in section 5 that the renormalization group equations describe the same
solution. If the reader is not interested in technical details, he can find the final
results in Eqs. (65) and (74), and can proceed to the next section.

There are only two possible dot states $s = 0, 1$. We denote by $\bar{s}$ the conjugate
state, $\bar{s} = 1(0)$ if $s = 0(1)$. Furthermore we define $\bar{p} = -p$ and $\bar{\mu} = -\eta r$ if $\mu = \eta r$.
The following three properties are needed for the following

$$\sum_{ps} (\overline{G_\mu^p})_{ss,\cdot\cdot} = 0 \quad , \quad \sum_{ps} p\,(\overline{I_\mu^p})_{ss,\cdot\cdot} = 0 \tag{62}$$

$$\sum_{pp'} (\overline{G_\mu^p \, G_{\mu'}^{p'}})_{s\bar{s},\cdot\cdot} = 0 \tag{63}$$

$$\lim_{D\to\infty} \sum_{pp'} (\overline{G_\mu^p G_{\mu'}^{p'}})_{s\bar{s},\cdot\cdot} = 0 \tag{64}$$

The time arguments of the interaction picture are not written explicitly. The
proof can be found in the appendix.

For the reduced density matrix $p(t)$ and the current $\langle I_L \rangle(t)$, we need the
kernels $\tilde{\Sigma}(z)$ and $\tilde{\Sigma}_I(z)$, see Eqs. (46) and (58). Due to particle number conservation of the total Hamiltonian, the reduced density matrix $p(t)$ stays diagonal
if the initial density matrix p_0 is diagonal. We therefore need only the diagonal
matrix elements $\tilde{\Sigma}(z)_{ss,s's'}$ and $\tilde{\Sigma}_I(z)_{ss,s's'}$ (the nondiagonal elements can also
be calculated but do not contribute to the current due to Tr_0 in Eq. (58)).

The kernels are defined in (42) and (56). Using property (62) we find

$$\boxed{\begin{aligned} \tilde{\Sigma}(z)_{ss,ss} &= -\tilde{\Sigma}(z)_{\bar{s}\bar{s},ss} \;, \\ \tilde{\Sigma}_I(z)_{ss,ss} &= \tilde{\Sigma}_I(z)_{\bar{s}\bar{s},ss} \;. \end{aligned}} \tag{65}$$

As a consequence, we only have to calculate the matrix element $(11, 00)$ of
the kernels (the matrix element $(00, 11)$ is analog and we quote the result at the
end).

We call an intermediate double-propagator state ss "even", and a state $s\bar{s}$
"odd". The vertices G and I change the parity. We want the matrix element
$(11, 00)$, i.e. we start with an even state 11 from the left. Thus, the state after
the first vertex of the kernels in Eqs. (42) and (56) is odd, and we can apply
properties (63) and (64) for the pair of the second and third vertex. We see that
the only possibility is that these two vertices are connected by a pair contraction.
After these two vertices we are again in an odd state and can proceed in the
same way. Finally we find that the only nonvanishing diagrams for $\tilde{\Sigma}(z)$ are

$$A\overset{\frown}{G}G \;\; \overset{\frown}{G}G \ldots \overset{\frown}{G}G\,B \;, \tag{66}$$

and analog for $\tilde{\Sigma}_I(z)$ by replacing $A \to I$.

We denote the sum over all sequences of $\overset{\frown}{GG}$-blocks in (66) by $\Pi(t)$, where t is the time difference between the boundary vertices A and B. $\Pi(t)$ can be calculated analogously to section 3.1 where we resummed sequences of Σ-blocks with the result (46) for the reduced density matrix in Laplace space. Thus we obtain

$$\tilde{\Pi}(z) = \frac{i}{z - L_0 - i\tilde{\sigma}(z)} , \tag{67}$$

with

$$\sigma(t) = \overset{\frown}{GG} = \gamma_{\mu\mu'}^{pp'}(t)G_\mu^p e^{-iL_0 t}G_{\mu'}^{p'} . \tag{68}$$

The kernels follow from

$$\Sigma(t) = \gamma_{\mu\mu'}^{pp'}(t)G_\mu^p \Pi(t)G_{\mu'}^{p'} , \tag{69}$$

$$\Sigma_I(t) = \gamma_{\mu\mu'}^{pp'}(t)I_\mu^p \Pi(t)G_{\mu'}^{p'} . \tag{70}$$

Using all definitions together with (26) and the fact that L_0 is a diagonal matrix, we find in the limit $D \to \infty$

$$\begin{aligned}
\sigma(t)_{10,10} &= -\gamma^-(-t) - \gamma^+(-t) = -\Gamma \delta(t) , \\
\tilde{\sigma}(z)_{10,10} &= -\Gamma/2 , \\
\tilde{\Pi}(z)_{10,10} &= \frac{i}{z - (L_0)_{10,10} - i\tilde{\sigma}(z)_{10,10}} = \frac{i}{z - \epsilon + i\Gamma/2} , \\
\Pi(t)_{10,10} &= e^{-i\epsilon t}e^{-\Gamma t/2} ,
\end{aligned} \tag{71}$$

with $\gamma^\eta = \sum_r \gamma_r^\eta$ and $\Gamma = \sum_r \Gamma_r$. Furthermore, using the symmetry relations (51) we get $\Pi(t)_{01,01} = \Pi(t)_{10,10}^*$. Using these results together with $\gamma_r^\eta(t)^* = \gamma_r^\eta(-t)$, we can evaluate (69) and (70), and find

$$\Sigma(t)_{11,00} = \gamma^+(t)\Pi(t)_{10,10} + \text{h.c.} = \gamma^+(t)e^{-i\epsilon t}e^{-\Gamma t/2} + \text{h.c.} \tag{72}$$

$$\Sigma_I(t)_{11,00} = (e/2)\gamma_L^+(t)\Pi(t)_{10,10} + \text{h.c.} = (e/2)\gamma_L^+(t)e^{-i\epsilon t}e^{-\Gamma t/2} + \text{h.c.} \tag{73}$$

Using (22) and (25) we finally get in Laplace space

$$\boxed{\begin{aligned}
\tilde{\Sigma}(z)_{11,00} &= \tfrac{i}{2\pi} \int dE \sum_r \Gamma_r f_r(E) \left\{ \tfrac{1}{E-\epsilon+z+i\Gamma/2} - \tfrac{1}{E-\epsilon-z-i\Gamma/2} \right\} \\
\tilde{\Sigma}_I(z)_{11,00} &= \tfrac{ie}{4\pi} \int dE \Gamma_L f_L(E) \left\{ \tfrac{1}{E-\epsilon+z+i\Gamma/2} - \tfrac{1}{E-\epsilon-z-i\Gamma/2} \right\}
\end{aligned}} \tag{74}$$

By an analog calculation one obtains the matrix element $(00, 11)$ by replacing $f_r \to 1 - f_r$ and changing the sign for the current kernel.

4   Renormalization Group

In this section we will develop a renormalization group technique to calculate the kernels $\tilde{\Sigma}(z)$ and $\tilde{\Sigma}_I(z)$ in a systematic way beyound perturbation theory. The two kernels are defined in (42) and (56). Except for the first boundary vertex, they are formally the same. Therefore, w.l.o.g. we discuss in the following the kernel $\tilde{\Sigma}(z)$. Furthermore, as pointed out after Eq. (42), we can set $t_0 = 0$.

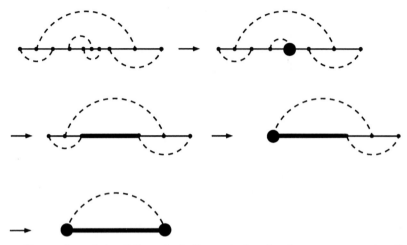

Fig. 4. Illustration of the RG-method. Successively, the shortest contraction line is integrated out. The thick lines and dots indicate renormalized propagators and vertices.

In section 2 we have found an effective theory in terms of the dot degrees of freedom. The reservoirs enter via pair contractions which couple the tunneling vertices. The aim is to integrate out all contractions in such a way that they can be interpreted as renormalization of L_0, G, A and B. The procedure is shown schematically in Fig. 4. We start with the integration over the shortest contraction. This gives rise to a renormalization of G in Fig. 4. The integration over the next shortest contraction renormalizes the propagator $e^{-iL_0(t_1-t_2)}$ of the dot. Proceeding in the same way, we find a renormalization of A and B in the next two steps. Finally we are left with a diagram which we can calculate easily by perturbation theory but with renormalized quantities. It can also happen that two or more vertices fall into one contraction which is integrated out, see e.g. Fig. 5. In this case we generate double-, triple-, and higher order vertices. In a perturbative renormalization group treatment one cuts this infinite hierarchy at a certain level. Here, we only consider propagator and single-vertex renormalization.

As described, we first want to integrate over the short contraction lines, i.e. over short time scales of the function $\gamma_\mu(t)$. Short time scales correspond to large energy scales. This leads us to the usual way renormalization group is formulated. One introduces a cut-off function $F(E/D)$ in the integrand of (21), where D is a

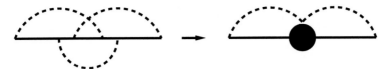

Fig. 5. Generation of a double-vertex.

high-energy cutoff, $F(0) = 1$ and $F(x)$ decays monotonically to zero sufficiently fast for $|x| \to \pm\infty$. As a consequence, the pair contraction $\gamma_{\mu,D}(t)$ depends on the cutoff D. In each renormalization group step one reduces the cutoff by an infinitesimal amount $D \to D - \delta D$, i.e. one tries to integrate out a small energy shell. The mathematical challenge is to interpret the result of this integration as a renormalization of system parameters, like L_0, G, A and B.

We prefer to develop the renormalization group procedure in real-time space. This turns out to be easier and more systematic. We introduce a cut-off function $F(t/t_c)$ which cuts off small time scales, i.e. $F(x)$, with $x > 0$, is a monotonically increasing function with $F(0) = 0$ and $\lim_{x \to \infty} F(x) = 1$, see Fig. 6. t_c is a cutoff parameter for small time scales. We define a cutoff dependent reservoir contraction by

$$\gamma_{\mu,t_c}(t) = \gamma_\mu(t)\, F(t/t_c) \;. \tag{75}$$

If we choose a sharp cutoff function $F(x) = \Theta(x-1)$, $\gamma_{\mu t_c}(t)$ includes only those time scales which are precisely larger than t_c. We note that the high-energy cutoff D introduced in Eq. (22) is independent of t_c and is not used as a renormalization group flow parameter. Within our real-time formulation, it corresponds to the physical band width of the reservoirs.

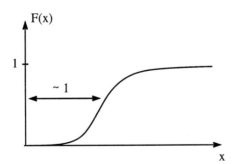

Fig. 6. The cutoff-function.

Motivated by the picture presented at the beginning of this section, we hope that all time scales being smaller than t_c, i.e. those being not present in $\gamma_{\mu t_c}(t)$ can be accounted for by renormalized quantities. To formulate this precisely let us write the kernel $\tilde{\Sigma}(z)$ as a functional of L_0, G, A, B and γ

$$\tilde{\Sigma}(z) = \mathcal{F}(L_0, G_\mu^p, A_\mu^p, B_\mu^p, \gamma_\mu(t)) \;, \tag{76}$$

where p, μ, and t run over all possible values within the functional. After replacing γ_μ by the cutoff-dependent function $\gamma_{\mu t_c}$ inside this functional, we try to find a cutoff dependence of all other quantities in such a way that the kernel stays invariant with the *same* functional $\mathcal{F}$

$$\tilde{\Sigma}(z) = \tilde{\Sigma}_{t_c}(z) + \mathcal{F}(L_{0t_c}, G^p_{\mu t_c}, A^p_{\mu t_c}, B^p_{\mu t_c}, \gamma_{\mu t_c}(t)) \ . \tag{77}$$

This equation is only exact if we neglect higher order vertex corrections, as already pointed out above. Otherwise we have to include all higher-order vertex terms as arguments in the functional $\mathcal{F}$ as well. Via $\gamma_{\mu t_c}$, the second term on the r.h.s includes only time scales which are larger than t_c. All other time scales are accounted for by the cutoff-dependence of L_0, G, A, and B. The first term on the r.h.s contains the contributions where *all* time scales are smaller than t_c. This part is not included in the second term where at least one contraction line $\gamma_{\mu t_c}$ occurs. It is important to notice that $\tilde{\Sigma}_{t_c}(z)$ should not be viewed as being trivially defined by Eq. (77). If we choose a sharp cutoff function $F(x) = \Theta(x-1)$, $\tilde{\Sigma}_{t_c}(z)$ can be contructively defined as containing only those diagrams where *all* contraction lines have a length smaller than t_c. For $t_c \to \infty$ the second term on the r.h.s. of (77) vanishes, provided the function $\gamma_\mu(t)$ decays sufficiently fast for $t \to \infty$. Thus, we obtain the final solution from

$$\tilde{\Sigma}(z) = \lim_{t_c \to \infty} \tilde{\Sigma}_{t_c}(z) \ . \tag{78}$$

The aim is to find differential equations which describe the cutoff-dependence of $\tilde{\Sigma}_{t_c}(z)$, L_{0t_c}, G_{t_c}, A_{t_c} and B_{t_c}. This will be the subject of the rest of this section.

Let us increase the cutoff by an infinitesimal amount $t_c \to t_c + dt_c$. Due to the invariance, Eq. (77) should not change

$$0 = d\tilde{\Sigma}(z) + \mathcal{F}(L_0 + dL_0, G + dG, A + dA, B + dB, \gamma + d\gamma) \ , \tag{79}$$

where we have omitted the subindex t_c and the indizes p and μ. The change of γ is known from the definition (75)

$$d\gamma_{\mu t_c}(t) = -\gamma_\mu(t) F'(t/t_c) t/t_c^2 \, dt_c \ , \tag{80}$$

or, for a sharp cutoff function $F(x) = \Theta(x - 1)$

$$d\gamma_{\mu t_c}(t) = -\gamma_\mu(t_c) \delta(t - t_c) \, dt_c \ . \tag{81}$$

The increment $d\gamma$ in (79) leads to all diagrams where one contraction line is replaced by $d\gamma$. We indicate this by a cross and define

$$\overset{\times}{\overline{G_1 \ldots G_2}} = - \left(\frac{d\gamma}{dt_c}\right)^{pp'}_{\mu\mu'}(t_1 - t_2) \, dt_c \, G^p_\mu(t_1) \ldots G^{p'}_{\mu'}(t_2) \ , \tag{82}$$

We call this a "cross contraction". For a sharp cutoff function, we get

$$\overset{\times}{\overline{G_1 \ldots G_2}} = \gamma^{pp'}_{\mu\mu'}(t_c) \delta(t_1 - t_2 - t_c) \, dt_c \, G^p_\mu(t_1) \ldots G^{p'}_{\mu'}(t_2) \ . \tag{83}$$

Corresponding definitions hold for G replaced by A or B. By convention we have included a minus sign into the definition since $-d\gamma$ accounts for the time scales of the contraction between t_c and $t_c + dt_c$, compare Eq. (81). We have to identify all terms created by $d\gamma$ with a contribution arising either from $d\tilde{\Sigma}$, dL_0, dG, dA or dB in Eq. (79) in order to fulfil invariance. Which one has to be taken depends on the number of vertices between G_1 and G_2, and wether one or both of the vertices are boundary vertices.

Let us start with the simplest case of two successive vertices which are not at the boundary

$$G_1 \overset{\frown{\times}}{G_2 G_3} G_4 \ . \tag{84}$$

Here, the two vertices G_1 and G_4 are contracted to any other vertices. Motivated by Fig. 4, it is tempting to take this diagram together with $G_1 G_4$ and interprete the cross contraction as a renormalization of L_0 in the following sense

$$G_1 G_4 \ + \ \int_{1>2>3>4} dt_2 dt_3 \, G_1 \overset{\frown{\times}}{G_2 G_3} G_4$$

$$\overset{?}{=} \ e^{iL_0 t_1} G^{p_1}_{\mu_1} e^{-i(L_0 + dL_0)(t_1 - t_4)} G^{p_4}_{\mu_4} e^{-iL_0 t_4} \ . \tag{85}$$

Expanding the exponential to linear order in dL_0, we see that this requires the identity

$$\int_{1>2>3>4} dt_2 dt_3 \, G_1 \overset{\frown{\times}}{G_2 G_3} G_4 \ \overset{?}{=} \ -i \int_{t_1>t>t_4} dt \, G_1 (dL_0)(t) G_4 \ . \tag{86}$$

To obtain such a relation we see immediately a problem. The integrand of the l.h.s contains a two-time object $\overset{\frown{\times}}{G_2 G_3}$, whereas the integrand of the r.h.s contains the single-time object $(dL_0)(t)$. We have to decide to which time-variable we identify t. Let us make an arbitrary choice: $t \equiv t_3$. Then it is obvious to try the ansatz

$$-i\,(dL_0)(t_3) \ = \ \int_{2>3} dt_2 \overset{\frown{\times}}{G_2 G_3} \ . \tag{87}$$

This is well-defined since it does not involve the time variables t_1 and t_4. However, let us try to insert this ansatz into the r.h.s of (86). We get two terms

$$-i \int_{1>3>4} dt_3 \, G_1 (dL_0)(t_3) G_4 = \int_{\substack{1>3>4\\2>3}} dt_2 dt_3 \, G_1 \overset{\frown{\times}}{G_2 G_3} G_4$$

$$= \int_{1>2>3>4} dt_2 dt_3 \, G_1 \overset{\frown{\times}}{G_2 G_3} G_4 \ + \ \int_{2>1>3>4} dt_2 dt_3 \, G_1 \overset{\frown{\times}}{G_2 G_3} G_4 \tag{88}$$

The first term agrees with the l.h.s. of (86), whereas the second one is a correction term which has to be considered separately. Let us try to interpret it as a term arising from dG. We define

$$(dG^{p_1}_{\mu_1})^{(1)}(t_1) = - \int_{2>1>3} dt_2 dt_3 \, G_1 \overset{\frown\!\ast\!\frown}{G_2 G_3} \,, \qquad (89)$$

where the subindex (1) indicates that this is the first contribution to dG (another one will follow below). Using this definition we finally get

$$\int_{1>2>3>4} dt_2 dt_3 \, G_1 \overset{\frown\!\ast\!\frown}{G_2 G_3} G_4 = -i \int_{1>3>4} dt_3 \, G_1 (dL_0)(t_3) G_4 + (dG^{p_1}_{\mu_1})^{(1)}(t_1) G_4$$

$$+ \int_{2>1>4>3} dt_2 dt_3 \, G_1 \overset{\frown\!\ast\!\frown}{G_2 G_3} G_4 \,. \qquad (90)$$

In the last term on the r.h.s. the two time-variables t_1 and t_4 are "clustered" together, they both have to lie within the time interval $[t_3, t_2]$. Therefore, we interpret this term as a double-vertex which we neglect. In conclusion, we have achieved our final goal: we have interpreted a term arising from $d\gamma$ by a renormalization of L_0 and G.

Let us try to understand what we have done so far and how to find a systematic procedure for obtaining the complete RG equations without going again into the details of the foregoing exercise. We have defined the renormalization of L_0 in (87), which is natural since this expression contains only operators of the dot. We write

$$-i\,(dL_0)(t_2) = \overset{\frown\!\ast\!\frown}{G_1 G_2} \,, \qquad (91)$$
$$\uparrow$$

where the arrow indicates the time variable which is not integrated out. It is *this* time variable which is used for the definition of the interaction picture of dL_0 and, most importantly, which is used for the definition of time-ordering, see the r.h.s of Eq. (86). The time-variable t_1 no longer appears explicitly because it is an internal integration variable within the definition of dL_0. This has the consequence, that time variables of other vertices do not care about the value of t_1 regarding time ordering. E.g., if we multiply $(dL_0)(t_3)$ with a vertex G_2 from the left, time-ordering only requires $t_2 > t_3$ but the ordering of t_2 with respect to the internal integration variable t_1 is not prescribed. Thus, due to dL_0, the following term will occur on the r.h.s. of (79)

$$\overset{\frown\!\ast\!\frown}{G_2 G_1 G_3} \,, \qquad (92)$$
$$\uparrow \quad \uparrow$$

where $t_1 > t_2 > t_3$. However, this term is not present in the original series because the ordering of operators does not agree with the ordering of time variables. Consequently, we have to subtract this term. This is the basic reason for the occurence of the second term on the r.h.s. of (90).

We interpret minus the correction term (92) as renormalization of G. Again we have to specify the time variable for the interaction picture and for time-ordering. We choose t_2 in analogy to Eq. (89). Together with the "real" vertex renormalization, arising from a free vertex inside a cross contraction, we get the complete renormalization of G

$$(dG^{p_2}_{\mu_2})(t_2) \;=\; \overbrace{G_1 \overset{\times}{G_2} G_3}^{} \; - \; \overbrace{G_2 \overset{\times}{G_1} G_3}^{} \;. \tag{93}$$

We can now proceed to look at other correction terms which can occur by multiplying (91) with two vertices from the left or (93) with one vertex from the left or right

$$\overbrace{G_2 G_3 \overset{\times}{G_1} G_4}^{} + \overbrace{G_2 \overset{\times}{G_1} G_3 G_4}^{} - \overbrace{G_2 G_3 \overset{\times}{G_1} G_4}^{} + \overbrace{G_1 \overset{\times}{G_2} G_4 G_3}^{} - \overbrace{G_2 \overset{\times}{G_1} G_4 G_3}^{} \;, \tag{94}$$

with $t_1 > t_2 > t_3 > t_4$. We have indicated the time-ordering variables of all sub-clusters which lead to these expressions. We see that the first and third term cancel each other. The other three correction terms have to be subtracted again. Minus the fifth term corresponds to the third term on the r.h.s. of Eq. (90). The second and fourth term are correction terms arising from the "real" vertex renormalization. However, all these correction terms correspond to double-vertices which we neglect consistently. Another double-vertex term arises from two free vertices within a cross contraction. Proceeding further in the same way, triple- and higher order vertex terms will occur which again are not considered here. We see that the procedure is very systematic and straightforward.

In the same way we can proceed for boundary vertices. Since we want to calculate the Laplace transform $\tilde{\Sigma}(z)$ of the kernel each diagram of the kernel (42) gets an additonal factor e^{izt}. Consequently we define the interaction picture for the boundary vertices slightly different from (41). We simply include the factor from the Laplace transformation

$$A^p_\mu(t) = e^{izt} A^p_\mu e^{-iL_0 t} \quad , \quad B^p_\mu(t) = e^{iL_0 t} B^p_\mu e^{-izt} \;. \tag{95}$$

Analog to the above procedure we can write down immediately all terms with cross contractions containing boundary operators. They can be interpreted as renormalization of $\tilde{\Sigma}$, A and B

$$d\tilde{\Sigma}(z) \;=\; \overbrace{A_1 \overset{\times}{B_2}}^{}\big|_{t_2=0} \;, \tag{96}$$

$$(dA^{p_2}_{\mu_2})(t_2) \;=\; \overbrace{A_1 \overset{\times}{G_2} G_3}^{} - \overbrace{A_2 \overset{\times}{G_1} G_3}^{} \;, \tag{97}$$

$$(dB^{p_2}_{\mu_2})(t_2) \;=\; \overbrace{G_1 \overset{\times}{G_2} B_3}^{} \;. \tag{98}$$

Again, we integrate implicitly over all time variables which are not indicated by an arrow, with $t_1 > t_2 > t_3$. Terms like $\overbrace{A_1 \overset{\times}{G_2}}^{}$ or $\overbrace{G_1 \overset{\times}{B_2}}^{}$ do not occur since they do not lead to connected diagrams. For this reason, there is also no correction term in the renormalization group equation for B.

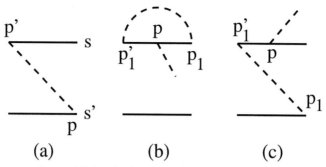

Fig. 7. Diagrams where additional minus occur if the ends of the indicated contraction are connected. In (a) a minus signs occurs if the particle number difference $N_s - N_{s'}$ is odd. In (b) and (c) a free vertex is crossed by connecting the end-points without crossing over p_0.

Eqs. (91), (93), and (96)-(98) are the final RG-equations. They can be translated easily to get explicit expressions. In order to get the renormalization of the bare quantities we have to set $t_2 = 0$ in all equations. To account for possible minus signs arising from commutation of fermionic reservoir field operators, we have to include two auxiliary sign functions. They arise because we have to connect the two end points of the cross contraction. If the cross contraction couples the upper with the lower propagator, a minus sign occurs if the intermediate double-propagator state at the vertex with the larger time is odd (i.e. if the difference of the fermionic particle numbers on the upper and lower propagator is odd), see Fig. 7a. We account this by a sign operator $\hat{\sigma}_\mu^{pp'}$ multiplying each cross contraction from the left. The matrix elements of this operator are defined by

$$(\hat{\sigma}_\mu^{pp'})_{ss',ss'} = \begin{cases} pp' & \text{for } N_s - N_{s'} = \text{odd} \\ 1 & \text{for } N_s - N_{s'} = \text{even} \end{cases}, \tag{99}$$

where N_s is the fermionic particle number of state s and μ is a fermionic contraction (for bosonic contractions no additional sign has to be considered). The second sign function concerns the case when one free vertex occurs within a cross contraction. A minus sign occurs when the cross contraction crosses over the free vertex when we want to connect the two end points. Since $\hat{\sigma}$ is multiplied from the left by definition, we connect the vertices always in such a way that we do not cross over p_0. We denote by p, p_1, and p'_1 the indices of the free vertex, the vertex with the larger and the one with the smaller time variable of the cross contraction, respectively. If $p = p'_1$ we get an additional minus sign, see Fig. 7b and 7c. Therefore we multiply the "real" vertex correction by an additional sign function

$$\eta_{\mu\mu'_1}^{pp'_1} = \begin{cases} -pp'_1 & \text{for } \mu \text{ and } \mu' \text{ fermionic} \\ 1 & \text{otherwise} \end{cases}. \tag{100}$$

In summary we get the following RG equations

$$\frac{d}{dt_c}\tilde{\Sigma}(z) = -\int\limits_0^\infty dt\,(\frac{d\gamma}{dt_c})^{pp'}_{\mu\mu'}(t)\hat{\sigma}^{pp'}_\mu A^p_\mu(t)B^{p'}_{\mu'}$$

$$\frac{d}{dt_c}L_0 = -i\int\limits_0^\infty dt\,(\frac{d\gamma}{dt_c})^{pp'}_{\mu\mu'}(t)\hat{\sigma}^{pp'}_\mu G^p_\mu(t)G^{p'}_{\mu'}$$

$$\frac{d}{dt_c}G^p_\mu = -\int\limits_0^\infty dt\int\limits_{-\infty}^0 dt'\,(\frac{d\gamma}{dt_c})^{p_1p'_1}_{\mu_1\mu'_1}(t-t')$$
$$\left[\eta^{pp'_1}_{\mu\mu'_1}\hat{\sigma}^{p_1p'_1}_{\mu_1}G^{p_1}_{\mu_1}(t)G^p_\mu - G^p_\mu\hat{\sigma}^{p_1p'_1}_{\mu_1}G^{p_1}_{\mu_1}(t)\right]G^{p'_1}_{\mu'_1}(t')$$

$$\frac{d}{dt_c}A^p_\mu = -\int\limits_0^\infty dt\int\limits_{-\infty}^0 dt'\,(\frac{d\gamma}{dt_c})^{p_1p'_1}_{\mu_1\mu'_1}(t-t')$$
$$\left[\eta^{pp'_1}_{\mu\mu'_1}\hat{\sigma}^{p_1p'_1}_{\mu_1}A^{p_1}_{\mu_1}(t)G^p_\mu - A^p_\mu\hat{\sigma}^{p_1p'_1}_{\mu_1}G^{p_1}_{\mu_1}(t)\right]G^{p'_1}_{\mu'_1}(t')$$

$$\frac{d}{dt_c}B^p_\mu = -\int\limits_0^\infty dt\int\limits_{-\infty}^0 dt'\,(\frac{d\gamma}{dt_c})^{p_1p'_1}_{\mu_1\mu'_1}(t-t')\,\eta^{pp'_1}_{\mu\mu'_1}\hat{\sigma}^{p_1p'_1}_{\mu_1}G^{p_1}_{\mu_1}(t)G^p_\mu B^{p'_1}_{\mu'_1}(t')$$

Implicitly we sum over all double indices on the r.h.s. of these equations which do not occur on the l.h.s.. We note that the overall sign on the r.h.s. differs from the one in [1] since we have included the factors $-i$ into the vertices. Furthermore, we have generalized the RG equations to an arbitrary cutoff-function $F(t/t_c)$. For a specific cutoff-function, e.g. $F(x) = \theta(x-1)$, and taking matrix elements of the RG-equations, all integrals in (4) can be calculated analytically. We are left with pure differential equations which can be solved numerically in a straightforward and very efficient way. Finally, the asymptotic value $\lim_{t_c \to \infty}\tilde{\Sigma}_{t_c}(z)$ gives the solution for the kernel, see Eq. (78). Using (46) and (47), we get the complete time-evolution of the reduced density matrix of the dot. The initial condition for the boundary operators at $t_c = 0$ is given by $A^p_\mu = B^p_\mu = G^p_\mu$. To get the current kernel $\tilde{\Sigma}_I(z)$, we take the same RG-equations with the initial conditions $A^p_\mu = I^p_\mu$ and $B^p_\mu = G^p_\mu$.

Numerically, one can not take the initial flow parameter $t^0_c = 0$ and the final one $t^f_c \to \infty$. However, one can check that the final solution is stabel for sufficiently small t^0_c and large t^f_c. This means that $1/t^0_c$ and $1/t^f_c$ have to be much larger resp. smaller than all other energy scales

$$\frac{1}{t^f_c} \ll \Gamma_L, \Gamma_R, T, eV, |\epsilon| \ll D \ll \frac{1}{t^0_c}\,, \tag{101}$$

where $eV = \mu_L - \mu_R$ is the bias voltage, and Γ_L, Γ_R are defined in (23). In addition, we want to send the band-width cutoff $D \to \infty$. This means that we have to check the stability of the solution for $t_c^0 \to 0$, $D \to \infty$, with $Dt_c^0 \to 0$. For the problem under consideration this is indeed possible.

There are three essential differences of our final RG equations to conventional poor man scaling and operator product expansion techniques [12,13]:

---

1. We have formulated the RG-equations within a real-time formalism on the Keldysh-contour.

2. We have not expanded the interaction picture of the vertex operators for small times, i.e. the renormalized propagation $\exp(\pm iL_0 t)$ is taken fully into account.

3. For $t_c > t_c^0$ we get $L_0 \neq [H_0, \cdot]$, i.e. we generate non-Hamiltonian dynamics during RG.

The first point is necessary for describing nonequilibrium phenomena. The second one is important for self-consistency reasons (L_0 in the exponent is a renormalized quantity) and for the stability of the solution for $t_c^f \to \infty$. Only if one expands the exponentials $\exp(\pm iL_0 t)$ in t, one has to stop the RG-flow when $|\lambda t_c| \sim 1$, where λ denotes any eigenvalue of L_0. We do not need this expansion and, consequently, are able to integrate out all time scales. The third property is the most important one. It is essential for a nonequilibrium theory to describe the physics of dissipation. Thus, the renormalized superoperator L_0 should no longer be expressible by a commutator with a renormalized Hamiltonian H_0, i.e. in matrix notation

$$(L_0)_{s_1 s_2, s_1' s_2'} \neq (H_0)_{s_1 s_2} \delta_{s_1' s_2'} - \delta_{s_1 s_2} (H_0)_{s_2' s_1'} . \tag{102}$$

To show this let us first define the renormalized Hamiltonian in a natural way. It is obvious that contractions which connect the upper with the lower propagator of the Keldysh contour, do not lead to Hamiltonian dynamics. This was already explained at the end of section 2.2. Thus, in order to define the renormalized H_0 let us consider the same RG equations but allow only for contractions within the upper or lower propagator. Formally, this means $p = p'$ and $p_1 = p_1'$ in (4). Under these conditions we can write the solution as

$$(L_0)_{s_1 s_2, s_1' s_2'} = (H_0)_{s_1 s_2} \delta_{s_1' s_2'} - \delta_{s_1 s_2} (H_0^\dagger)_{s_2' s_1'} . \tag{103}$$

To obtain this, we have used the symmetry relations (51) together with the property

$$\gamma_{\mu \mu'}^{p p'}(t)^* = \gamma_{\mu \mu'}^{p p'}(-t) = \gamma_{\bar\mu \bar\mu'}^{\bar p \bar p'}(t) . \tag{104}$$

The renormalized Hamiltonian H_0 is defined by considering only the upper propagator, i.e. setting $p = p' = p_1 = p'_1 = +$ and replacing $L_0 \to H_0$ and $G^+_\mu \to -ig_\mu$ in the RG-equations. We note that H_0 is non-hermitian since we generate complex energies during RG. Physically this describes broadening of levels or finite life-times. From (103) we conclude that the backward propagator involves $H^\dagger_0$, i.e. the sign of energy-broadening is different for the forward and backward propagator. This means, that even if we disregard contractions connecting the upper with the lower propagator, we get $L_0 \neq [H_0, \cdot]$. However, this does not imply the physics of dissipation. It describes the physics of a finite broadening of the energy levels due to the coupling to the environment. This reflects the Heisenberg-uncertainty relationship.

In contrast, if we couple the forward and backward propagator by contractions, the matrix L_0 will completely change its form. It can neither be represented as (102) nor as (103). The coupling of the propagators gives rise to the generation of rates where the states on the upper and lower propagator are changed simultaneously. Rates describe the evolution of the system into a stationary state, i.e. we generate irreversibility or dissipation during RG.

The RG equations preserve conservation of probability and give real expectation values for hermitian observables. Once the symmetry relations and sum rules, stated in Eqs. (51)-(54) and (60)-(61), are fulfilled initially, they are not changed by the RG-flow. Thus, by applying the same proof as in section 3.1, we obtain a normalized probability distribution of the dot and a real expectation value for the current.

Under certain circumstances the RG equations can be simplified. This will be important for the exact solution presented in the next section. We decompose the pair contraction (20) trivially into two terms

$$\gamma_\mu(t) = \gamma^\delta_\mu(t) + \bar{\gamma}_\mu(t) , \qquad (105)$$

with

$$\gamma^\delta_\mu = \frac{1}{2}\langle [j_{\bar{\mu}}(t), j_\mu]_{-\sigma}\rangle \quad , \quad \bar{\gamma}_\mu = \frac{1}{2}\langle [j_{\bar{\mu}}(t), j_\mu]_\sigma\rangle , \qquad (106)$$

where $\sigma = \pm$ for $\mu =$ bosonic (fermionic), and $[\cdot, \cdot]_{-\sigma}$ denotes the (anti-)commutator for $\sigma = \pm$. This decomposition is useful since in many problems, the part $\gamma^\delta_\mu(t)$ turns out to be proportional to a delta function or derivatives of a delta function (at least for the band width cutoff $D \to \infty$). For our special model we get (compare (26))

$$\gamma^\delta_\mu(t) = \frac{\Gamma_r}{2}\,\delta(t) . \qquad (107)$$

Such a contribution can be included into the initial conditions of the RG equations. First, we decompose $\gamma^{pp'}_{\mu\mu'}(t) = \gamma^{pp'}_{\mu\mu'}(t)^\delta + \bar{\gamma}^{pp'}_{\mu\mu'}(t)$ as well by inserting (105) into (30)

$$\gamma^{pp'}_{\bar{\mu}\mu}(t) = [p'\gamma^\delta_\mu(t) + \bar{\gamma}_\mu(t)] \begin{cases} 1 \text{ for } \mu = \text{ bosonic} \\ p' \text{ for } \mu = \text{ fermionic} \end{cases} . \qquad (108)$$

We have obtained the explicit dependence on the indices p and p' in this equation. Therefore we try to include these factors into the definition of the vertex operators. Having included the γ_μ^δ-parts into the initial conditions, it turns out that closed RG equations can be found for the quantities

$$(G_\mu)_{ss',\cdot\cdot} = \begin{cases} \sum_p (G_\mu^p)_{ss',\cdot\cdot} & \text{for } N_s - N_{s'} = \text{ even or } \mu = \text{ bosonic} \\ -i \sum_p p (G_\mu^p)_{ss',\cdot\cdot} & \text{otherwise} \end{cases},$$

(109)

with N_s being the fermionic particle number of state s. In the same way we define the boundary vertices A_μ and B_μ. We have included a factor $-i$ for the second case in (109) in order to get the same symmetry relations as in (51)

$$(G_\mu)^*_{s_1 s'_1, s_2 s'_2} = (G_{\bar\mu})_{s'_1 s_1, s'_2 s_2} .$$

(110)

Using the fact that fermionic vertex operators $(G_\mu)_{ss', s_1 s'_1}$ change the parity of the fermionic particle number difference $N_{s_1} - N_{s'_1} \to N_s - N_{s'}$, we get after a lengthy but straightforward calculation the following RG-equations arising from the $\tilde\gamma_\mu$-parts

$$\frac{d}{dt_c}\tilde\Sigma(z) = -\int_0^\infty dt\, \frac{d\tilde\gamma_\mu}{dt_c}(t) A\bar\mu(t) B_\mu$$

$$\frac{d}{dt_c}L_0 = -i\int_0^\infty dt\, \frac{d\tilde\gamma_\mu}{dt_c}(t) G_{\bar\mu}(t) G_\mu$$

$$\frac{d}{dt_c}G_\mu = -\int_0^\infty dt \int_{-\infty}^0 dt'\, \frac{d\tilde\gamma_{\mu_1}}{dt_c}(t - t')\left[\sigma_{\mu\mu_1} G_{\bar\mu_1}(t) G_\mu - G_\mu G_{\bar\mu_1}(t)\right] G_{\mu_1}(t')$$

$$\frac{d}{dt_c}A_\mu = -\int_0^\infty dt \int_{-\infty}^0 dt'\, \frac{d\tilde\gamma_{\mu_1}}{dt_c}(t - t')\left[\sigma_{\mu\mu_1} A_{\bar\mu_1}(t) G_\mu - A_\mu G_{\bar\mu_1}(t)\right] G_{\mu_1}(t')$$

$$\frac{d}{dt_c}B_\mu = -\int_0^\infty dt \int_{-\infty}^0 dt'\, \frac{d\tilde\gamma_{\mu_1}}{dt_c}(t - t')\, \sigma_{\mu\mu_1} G_{\bar\mu_1}(t) G_\mu B_{\mu_1}(t')$$

(111)

where

$$\tilde\gamma_\mu(t) = \bar\gamma_\mu(t) \begin{cases} 1 & \text{for } \mu = \text{ bosonic} \\ i & \text{for } \mu = \text{ fermionic} \end{cases},$$

(112)

and

$$\sigma_{\mu\mu'} = \begin{cases} 1 & \text{for } \mu \text{ or } \mu' = \text{ bosonic} \\ -1 & \text{for } \mu \text{ and } \mu' = \text{ fermionic} \end{cases}.$$

(113)

We see that the new RG equations are more compact but we note that they can not be applied to all models. Sometimes, like e.g. in spin boson models, it happens that the part $\gamma_\mu^\delta(t)$ can not be expressed by the derivative of a delta function and, consequently, can not be incorporated into the initial conditions. In this case, one should use the original and generally valid RG-equations (4).

5   Exact Solution of the RG Equations

In this section we demonstate that the RG equations can be solved exactly for the special model (1)-(3) of a single non-degenerate dot state. It turns out that the result for the kernels agrees with that of section 3.2. We conclude that the RG-approach solves the present model exactly.

We choose the sharp cutoff-function $F(x) = \theta(x - 1)$ and get from the RG-equation (4) for a matrix element of the kernel

$$\tilde{\Sigma}(z)_{ss,\bar{s}\bar{s}} = \sum_{pp'\mu s'} \int_0^\infty dt_c \ \gamma_{\bar{\mu}\mu}^{pp'}(t_c) \, e^{izt_c} \, (A_{\bar{\mu}}^p)_{ss,s'\bar{s}'} \, e^{-i\alpha_{s'}t_c} \, (B_\mu^{p'})_{s'\bar{s}',\bar{s}\bar{s}} \ . \quad (114)$$

Here, due to particle number conservation, we have used the fact that the only nonvanishing matrix elements of L_0 are $\alpha_s = (L_0)_{s\bar{s},s\bar{s}}$ and $(L_0)_{ss,s's'}$. To evaluate this equation we need the t_c dependence of α_s and the boundary vertex operators. To obtain this it is convenient to include the part γ_μ^δ into the initial conditions. The latter follow from inserting the part $\gamma_{\bar{\mu}\mu}^{pp'}(t_c)^\delta$ into the RG-equations (4), and integrating. We denote the initial value for L_0 by L_0^δ. The vertex operators do not obtain any initial renormalization from the δ-function parts since there is no phase space for the time integrals in (4). Taking the part $\gamma_{\bar{\mu}\mu}^{pp'}(t_c)^\delta$ from the first term of (108) together with (107), and inserting it in the RG-equation (4) for L_0, we find

$$(\frac{d}{dt_c})\alpha_s^\delta = i\frac{\Gamma}{2}\delta(t_c)pp'(G_{\bar{\mu}}^p G_\mu^{p'})_{s\bar{s},s\bar{s}} = -i\Gamma\delta(t_c) \ . \quad (115)$$

This gives

$$\alpha_0^\delta = -\epsilon - i\frac{\Gamma}{2} \quad , \quad \alpha_1^\delta = \epsilon - i\frac{\Gamma}{2} \ , \quad (116)$$

where the part with the single particle energy ϵ stems from the initial value without any renormalization.

As we will show in the following we find no further renormalization from the $\tilde{\gamma}_\mu$-part, i.e. using the simplified set of RG-equations (111), we get

$$\frac{d}{dt_c}\alpha_s = \frac{d}{dt_c}(A_\mu)_{ss,s'\bar{s}'} = \frac{d}{dt_c}(B_\mu)_{s'\bar{s}',\bar{s}\bar{s}} = \frac{d}{dt_c}G_\mu = 0 \ . \quad (117)$$

This means that the renormalization of the kernel due to the $\tilde{\gamma}_\mu$-part can be calculated with unrenormalized vertex operators and using the result (116) for

α_s. Since, trivially, the same applies to the γ_μ^δ-part, we can directly evaluate (114) in this way and get the result (74) of section 3.2.

What remains to be shown is Eq. (117). We first note the following properties of the matrix elements of the vertex operator, which follow from the definition (109)

$$\sum_s (G_\mu)_{ss,s'\bar{s}'} = 0 \quad , \quad (G_\mu)_{s\bar{s},s's'} = (G_\mu)_{s\bar{s},\bar{s}'\bar{s}'} \ . \tag{118}$$

Furthermore, using (53), we get

$$\sum_s (e^{-iL_o t})_{ss,\cdot\cdot} = 1 \ . \tag{119}$$

We apply these properties to the following expression

$$[G_\mu(t)G_{\mu'}(t')]_{s\bar{s},s'\bar{s}'} =$$

$$= e^{i\alpha_s t} e^{-i\alpha_s' t'} \sum_{s_1 s_2} (G_\mu)_{s\bar{s},s_1 s_1} (e^{-iL_0(t-t')})_{s_1 s_1, s_2 s_2} (G_{\mu'})_{s_2 s_2, s'\bar{s}'}$$

$$= e^{i\alpha_s t} e^{-i\alpha_s' t'} (G_\mu)_{s\bar{s},s_1 s_1} \sum_{s_2} (G_{\mu'})_{s_2 s_2, s'\bar{s}'} \quad = 0 \ . \tag{120}$$

Using this result after taking matrix elements of the RG-equations (111), we find immediately (117).

We conclude that after having set up the general RG-equations (4) and (111), the exact solution of the present model can be found analytically in a straightforward way. There is no need to consider diagrammatic details as in section 3.2. Furthermore, we see that the effect of the reservoirs is simply a broadening of the local state of the dot by $\Gamma/2$, see (116). Otherwise the evaluation of the kernel is the same as in lowest order perturbation theory in Γ.

6   Summary and Outlook

In this article we have presented a new viewpoint to analyse nonperturbative aspects of nonequilibrium systems. We considered a small system coupled linearly to several baths. Nonperturbative means that the coupling between system and bath is so strong that quantum fluctuations induce broadening of the states in the system together with possible renormalization of energy levels and the coupling to the environment. In macroscopic systems, such effects are negligible since the interaction with the environment is a surface effect. In contrast, our goal is to describe mesoscopic systems like quantum dots, magnetic nanoparticles or chemical molecules coupled to particle or heat reservoirs. Usually such systems are treated within perturbation theory [14–16]. This means that the kernel $\Sigma(t)$ of the kinetic equation is calculated in lowest order perturbation theory in the coupling, the so-called golden rule or Pauli-Master equation approach. Our aim here was to find a systematic way to consider self-consistently an infinite series of higher-order contributions to the kernel. This is a nontrivial task, especially in nonequilibrium where a real-time formalism on a Keldysh contour has to be used. Renormalization effects known from equilibrium theories have to be incorporated consistently within a kinetic equation. Our point of view relies on renormalization group ideas. We try to integrate out all energy scales of the bath in infinitesimal steps. Each step is interpreted as a change of various quantities, like broadening and renormalization of energy levels, and generation of rates. For formal reasons we have set up this procedure in real-time space. During this procedure we keep the kernel of the kinetic equation invariant and generate non-Hamiltonian dynamics to describe the physics of dissipation. The final RG-equations are presented in (4), and, alternatively, in (111). Solving them, provides the complete description of the time-evolution of the reduced density matrix of the system and arbitrary observables being linear in the field operators of the bath. This includes the consideration of an initally out of equilibrium state as well as the description of stationary nonequilibrium situations. We have solved the equations exactly for the special case of a quantum dot with one state coupled to two particle reservoirs. The solution turned out to be identical to the exact one.

The reader might argue that this is a trivial result since the model under consideration is very simple. However, as shown in a recent paper, the same RG equations provide not only the exact solution for a noninteracting quantum dot with one state, but gives a good solution also for the spin-degenerate case including a finite on-site Coulomb interaction U [1], the so-called Anderson impurity model in nonequilibrium. Here, not only energy broadening but also energy and coupling constant renormalizations are important. In the mixed-valence and empty-orbital regime, i.e. for level positions near or above the Fermi level of the reservoirs, it was shown that the linear conductance and the average occupation in equilibrium agrees perfectly with Friedel sum rules, Bethe ansatz, and numerical renormalization group methods within 2-3%. This means that the RG provides a good solution for the whole range from $U = 0$ to $U = \infty$. This is a surprising result since usually methods designed for strong interaction do not work well for weak interaction and vice versa. Therefore, the fact that the RG gives the exact result for the noninteracting case is not at all a trivial result. In fact, within the slave boson technique [17], which is a well-known and well-established method in the theory of strongly correlated Fermi systems, it is very complicated to find the exact solution for $U = 0$ [17].

Another example where the RG-method has been applied is the study of transport through the metallic single-electron transistor [18]. For this case, the RG-equations have been solved in sixth-order perturbation theory in the coupling. Surprisingly, it turned out that the solution agrees very well with exact perturbation theory in the same order. The reason why the RG-equations have not been solved in all orders is that the result was not finite for the band width cutoff $D \to \infty$. Only if certain parts of double-vertex corrections were included, the solution turned out to be cutoff-independent. A more detailed study of the D-dependence and the solution for higher orders is currently under way.

The RG-equations on the level of propagator and single-vertex renormalization are pure differential equations and, consequently, can be solved numerically very quickly. Typical response times range from seconds to a few minutes for one set of parameters, even on a usual PC maschine. There are still some problems with the asymptotic solution of the RG equations for $t_c \to \infty$. First, it is not known rigorously wether a stationary solution exists at all. Secondly, the numerical solution is plagued by oscillating functions, typical for real-time problems. However, for the problems studied so far, and provided the coupling is not too strong, there is a stationary numerical solution over a sufficiently long period in t_c. This applies at least to the physical quantities under consideration, like the probability distribution and the current.

Another reason for the efficiency of the present method is the possibility to calculate physical observables directly without the need of correlation functions like in linear response theory. Furthermore, if desired, correlation functions can also be studied with the RG-method. In this case, there are additional RG-equations to describe the renormalizations of the external vertices defining the correlation function. These equations together with explicit solutions will be published in forthcoming works.

The method is also applicable to the study of the ground state energy since the RG-flow on the single forward propagator provides the S-matrix. This idea has been applied to the single-electron box [10], coupled metallic islands [19], and the one-dimensional Polaron problem [20]. In the first two cases very good results have been obtained, even comparable to very time-consuming QMC-simulations. For the 1d-Polaron problem, the results were at least satisfactory for small coupling. Here, problems occured since the correlation function of the bath does not decay for long times and undamped modes occur which make the numerical analysis of the asymptotic solution very difficult.

Finally, we remark that a challenge for future research is the consideration of higher-order vertex corrections. Our RG-scheme provides a systematic treatment for setting up RG-equations for all kinds of multiple vertices. However, even on the level of double-vertices, the number of terms increases considerably and the RG-equations become integral-differential equations. The reason is the retarded nature of double-vertices. Therefore it is necessary to find physical arguments to select the most important terms or to improve the numerical efficiency by neglecting the retardation in a convenient way. Whereas single-vertices describe basically charge fluctuations (in case of coupling to particle reservoirs), double-vertices describe physical processes via virtual intermediate states where the local system can change its state without changing its particle number. This is e.g. important for the study of spin fluctuations in local impurities or quantum dots, leading to the Kondo effect [8,12]. Although such processes are perturbatively included in the RG-equations set up in this article, it is important to study them fully self-consistently by considering corresponding RG-equations for double-vertex terms on a Keldysh contour.

Acknowledgements

Useful discussions are acknowledged with T. Costi, J.v. Delft, H. Grabert, G. Schön, K. Schönhammer, and P. Wölfle. I owe special thanks to J. König, F. Kuczera, T. Pohjola, and M. Keil for very fruitful collaborations on applying the RG-method to various systems. This article was supported by the Swiss National Science Foundation (H.S.) and by the "Deutsche Forschungsgemeinschaft" as part of "SFB 195".

Appendix

Here we prove Eqs.(62)-(64). The first one follows from

$$(\overset{\frown}{G_\mu^+})_{ss,\bar{s}s} + (\overset{\frown}{G_\mu^-})_{\bar{s}\bar{s},\bar{s}s} = 0 \ , \tag{121}$$

which can easily be seen by using the definitions (27) and (28). The reservoir contraction is the same in both terms since G_μ^p is the vertex at later time, compare (31). The same proof can be used for I_μ^p by using the definition (37). Since there is no change of sign of I_μ^p if p is changed to $\bar{p}$, we have to add a factor p under the sum in (62).

To show the second property (63), we again use the definitions (27) and (28), and find

$$(\overset{\frown}{G_\mu^p \overset{\frown}{G_{\mu'}^{p'}}})_{s\bar{s},\cdots} + (\overset{\frown}{G_\mu^{\bar{p}} \overset{\frown}{G_{\mu'}^{\bar{p}'}}})_{s\bar{s},\cdots} = 0 \ . \tag{122}$$

Both reservoir contraction are the same in both terms since both vertices are at later time. However, the reader can convince himself very easily that there is an additional relative sign between the two terms of (122) due to the interchange of reservoir Fermi operators.

The third property (64) follows from

$$\lim_{D\to\infty} \{(\overset{\frown}{G_\mu^p \overset{\frown}{G_{\mu'}^{p'}}})_{s\bar{s},\cdots} + (\overset{\frown}{G_\mu^{\bar{p}} \overset{\frown}{G_{\mu'}^{\bar{p}'}}})_{s\bar{s},\cdots}\} = 0 \ . \tag{123}$$

Here, by comparing the two terms, the contraction associated with the second vertex changes from γ_r^η to $\gamma_r^{-\eta}$. Otherwise the two terms are the same. However, the sum $\gamma_r^\eta(t) + \gamma_r^{-\eta}(t)$ gives a $\delta(t)$-function in the limit $D \to \infty$, see (26). This means that there is no phase space in time for the first vertex and the sum is zero.

References

1. H. Schoeller and J. König, cond-mat/9908404.
2. A.L. Fetter and J.D. Walecka, *"Quantum Theory of Many Particle Systems"*, (McGraw-Hill, New York, 1971).
3. P.W. Anderson, J. Phys. C **3**, 2346 (1970); K.G. Wilson, Rev. Mod. Phys. **47**, 773 (1975).

4. H.R. Krishna-murthy, J.W. Wilkins, and K.G. Wilson, Phys. Rev. B **21**, 1003 (1980); Phys. Rev. **21**, 1043 (1980).
5. T.A. Costi, A.C. Hewson, and V. Zlatić, J. Phys. C **6**, 2519 (1994).
6. S.D. Glazek and K.G. Wilson, Phys. Rev. D **48**, 5863 (1993); F. Wegner, Ann. Physik (Leipzig), **3**, 77 (1994).
7. P. Joyez, V. Bouchiat, D. Esteve, C. Urbina, and M.H. Devoret, Phys. Rev. Lett. **79**, 1349 (1997).
8. D. Goldhaber-Gordon *et al.*, Nature (London), **391**, 156 (1998); Phys. Rev. Lett. **81**, 5225 (1998); S.M. Cronenwett *et al.*, Science **281**, 540 (1998); J. Schmid *et al.*, Physica B **256-258**, 182 (1998); F. Simmel *et al.*, Phys. Rev. Lett. **83**, 804 (1999).
9. H. Schoeller, in *'Mesoscopic Electron Transport'*, eds. L.L. Sohn *et al.* (Kluwer 1997), p. 291; J. König, "Quantum Fluctuations in the Single-Electron Transistor", Shaker (Aachen, 1999).
10. J. König and H. Schoeller, Phys. Rev. Lett. **81**, 3511 (1998).
11. C. Caroli et al., J. Phys. C **4**,916 (1971); J. Phys. C **5**, 21 (1972).
12. A.C. Hewson, *The Kondo Problem to Heavy Fermions*, (Cambridge University Press, 1993).
13. I. Affleck and A.W.W. Ludwig, Nucl. Phys. B **360**, 641 (1991).
14. K. Blum, *Density Matrix Theory and Applications*, 2nd edition (Plenum Press, 1996).
15. E. Fick and G. Sauermann, *The Quantum Statistics of Dynamic Processes*, (Springer, 1990).
16. C.W. Gardiner, *Quantum noise*, Springer Series in Synergetics 56, (Springer-Verlag, Berlin, 1991).
17. S.E. Barnes, J. Phys. F: Metal Phys. **6**, 1375 (1976); **7**, 2637 (1977).
18. F. Kuczera, H. Schoeller, J. König, and G. Schön, submitted to Europhys. Lett..
19. T. Pohjola, J. König, H. Schoeller, and G. Schön, Phys. Rev. B **59**, 7579 (1999).
20. M. Keil and H. Schoeller, preprint 1999.

Spin States and Transport in Correlated Electron Systems

Hideo Aoki

Department of Physics, University of Tokyo, Hongo, Tokyo 113-0033, Japan

Abstract. We explore theoretically how the correlation effects involving spins can appear in transport properties when we put zero dimensional (0D: quantum dots), 1D (quantum wires) or in 2D correlated systems in magnetic fields. We show that the correlation induces a spin blockade through 'electron molecules' in quantum dots, a spin polarised current in a Tomonaga-Luttinger liquid and a negative magnetoresistance in partly flat bands,

1 Introduction

Recent studies of mesoscopic systems have brought to light many unusual features in their quantum transport properties [1]. This is heightened by recent advances in fabricating low-dimensional structures, i.e., quantum dots, quantum wires or quasi-two-dimensional crystal structures.

The physics of strongly correlated electron systems is another landmark of these decades, and fascination of electron correlation phenomena ever increases. In particular, it is becoming increasingly clear that the spin degrees of freedom play a crucial role in strongly correlated electron systems, since the way in which electron correlation effects manifest themselves is dominated by the spin state. It is now generally recognised that spins do indeed dominate in ordinary correlated electron systems such as the Hubbard model, often evoked to model high-T_C cuprates [2], or unusual correlated systems such as the fractional quantum Hall system [3].

Involvement of spins must appear in transport properties in correlated electron systems. The purpose of the present article is to propose novel phenomena along this line, by posing in particular a question: what will happen if we put various correlated electron systems in magnetic fields, B (Fig.1)?

Here we show that we do have intriguing phenomena in various spatial dimensions(D) as

- 0D+B: spin blockade in quantum dots [4],
- 1D+B: spin polarised current in quantum wires [5],
- 2D+B: magnetoresistance in flat bands [6].

(i) in a quantum dot containing few electrons the electron correlation enforces the electrons to take what may be called an 'electron molecule' [11,12] comprising Landau's orbits in strong magnetic fields. Thus, while in ordinary molecules there are heavy nuclei around which electrons are bound and we start making Born-Oppenheimer approximations, etc, we have here no nuclei and electrons just

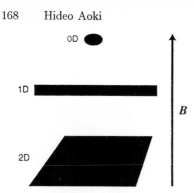

Fig. 1. Correlated electron systems in 0D (quantum dot), 1D (quantum wire) and 2D in a magnetic field, **B**.

spontaneously form molecules due to the repulsive interaction. A combination of the molecular symmetry and Pauli's exclusion principle then results in a specific series ('magic number') of the total angular momentum. As the magnetic field is varied the total angular momentum jumps from one magic number to another, where the total spin of the system, dominating how the magic states appear, jumps in a wild manner. This can be used to realise a spin blockade, where electrons are prohibited to tunnel due to a selection rule in the spin wavefunction.

(ii) When a 1D quantum wire, which can universally described with Tomonaga-Luttinger (TL) model, is put in a magnetic field, we have a two-component TL system with different Fermi velocities for different spin directions. From this we predict that the ratio of the conductivity for the up spin to that for the down spin becomes very large at low temperatures, which will result in a spin polarised current.

(iii) When a system, in any dimension, has a band dispersion which contains a flat part around the Fermi energy: magnetoresistance in flat bands. The repulsive electron correlation should cause ferromagnetic spin fluctuations to develop with an enhanced susceptibility. A relatively small magnetic field will then shift the majority-spin Fermi level to the dispersive part, which gives a novel mechanism for a negative magnetoresistance. This idea is analytically confirmed by calculating the conductivity in magnetic fields with the Hubbard model for a lattice representing an experimentally examined organic conductor,

2   Spin Blockade in Quantum Dots

2.1   Introduction

The Coulomb blockade is one of the highlights in the transport properties of mesoscopic systems. This is a phenomenon where a resonant tunnelling from an electrode into a dot cannot occur unless the chemical potential of the electrode coincides with one of the discrete energy levels of the finite system + the electric charging energy. While this has to do with the charge degrees of freedom, Weinmann *et al.* have recently proposed to use spin degrees of freedom, where the dot should be blocked, at zero temperature, due to the spin selection rule in the overlap of spin wavefunctions, $\langle \psi_{\text{spin}}^{\text{final}} | \psi_{\text{spin}}^{\text{initial}} \rangle$, that involves Clebsch-Gordan

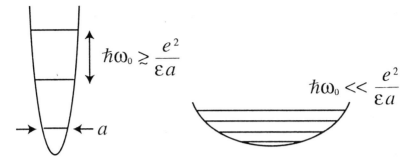

weak interaction   strong interaction

Fig. 2. Weak- and strong-interaction regimes for quantum dots are schematically depicted, where $\hbar\omega_0$ is the one-electron level spacing in the confining potential while $e^2/\epsilon a$ is the typical Coulomb repulsion in the dot.

coefficients if the total spins of the ground state of the initial state (with $(N-1)$ electrons) and the final (N) differ by more than $1/2$ [7,8]. This is called the 'spin-blockade' and has been studied theoretically for weak electron interaction regimes. When the electron-electron interaction is weak, i.e., when the spacing, $\hbar\omega_0$, between the discrete one-electron levels in the dot is greater than the Coulombic energy, $e^2/\epsilon a(a$: size of the dot, ϵ the dielectric constant) (Fig.2), we can take a Hund's coupling picture, in which electrons are accommodated in one-electron states with high spins for degenerate states. Since the spin changes only by $1/2$ in this rule the spin-blockade condition is not realised, so that some modifications such as an anharmonicity in the confinement potential [9] have to be introduced.

We propose here that the situation should be dramatically altered when we put quantum dots in strong magnetic fields [4]. Then the ground states change into the magic-number states [10,11]. This remarkable effect comes from the electron correlation effect, where the total angular momentum has to take a specific (magic number) values due to a combined effect of the 'molecular' configuration of electrons (triangular for three electrons, square for four, etc) and Pauli's principle [12–14]. The total angular momentum of the ground state jumps from one magic number to another as the magnetic field **B** is varied. The Coulomb interaction is strong in small dots even for $B = 0$ where we have $\hbar\omega_0(\sim$ few meV$) \gg e^2/\epsilon a$, but a crucial point here is that an application of a magnetic field pushes the situation to the strong-correlation side much further.

A tell-tale indication that electron correlation is really at work is the fact that the total spin (S in $\mathbf{S}^2 = S(S+1)$), not to be confused with S_z) of the ground state, which dominates how the electrons correlate, changes wildly as shown in Fig. 6 below. This happens when the typical Coulomb energy is much greater than the single-electron level spacing, where electron molecule are formed. In this sense this is genuinely an electron-correlation effect.

What we have found is that this can be utilised to realise a spin blockade. We have numerically studied the ground and excited states of a dot that contain two to four electrons with a parabolic confinement potential and find that the spin blockade should indeed be observed. Physically, we can interpret the result in terms of the 'electron molecules' [11,12,14], which enables us to regard the spin wavefunctions taking part in the spin blockade as spin configurations including the resonating valence bond (RVB) states, that are usually invoked for molecules and lattice systems. We further show that the spin-blockade condition is easier to satisfy in double dots which can be tuned by controlling the layer separation and the strength of the interlayer tunnelling.

2.2   Magic Numbers

So let us start with looking at the total angular momentum (L) of 2D electrons confined in a quantum dot in a magnetic field.

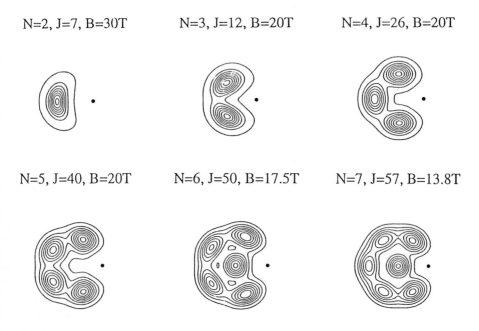

N=2, J=7, B=30T   N=3, J=12, B=20T   N=4, J=26, B=20T

N=5, J=40, B=20T   N=6, J=50, B=17.5T   N=7, J=57, B=13.8T

Fig. 3. Quantum mechanical density correlation with one electron fixed at the position indicated by a filled circle for typical magic-number states in quantum dots[12].

The numerical result (Fig.4) [12] shows that the total Coulombic energy of the interacting electrons, although roughly a decreasing function of L as the electrons move further apart for larger L, has a series of downward cusps. The presence of such magic L values signifies that the ground state with the minimum the total energy (Landau's energy in the presence of the confinement + Coulombic energy) has L values that jumps from one magic number to another as B is

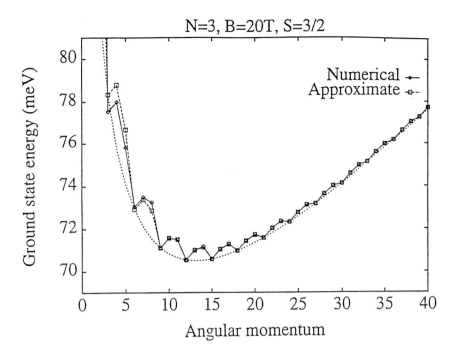

Fig. 4. The lowest-energy against the total angular momentum for a three-electron dot. The full line is the exact diagonalisation result, while the dashed line, almost indistinguishable from the full line, is the electron-molecule result[12].

increased. Since L has a one-to-one correspondence with the spatial size ($\propto \sqrt{L}$) of the wavefunction, the size is not a smooth function of B, but undergoes a series of jumps (overall a shrinking function of B, but expands at a transition between the magic numbers [15,16]. For example, the total angular momentum L of three spin-polarized electrons changes $3 \to 6 \to 9 \to \dots$ with increasing magnetic field.

Maksym has explained this as an effect of correlation in the electron configuration, where Pauli's exclusion principle dictates group-theoretically the manner in which the quantum numbers should appear [12]. There, the picture of the 'electron molecule', in which the electrons with a specific configuration (triangle for three electrons, square for four, etc) rotating as a whole has turned out to be surprisingly accurate. This continues to be the case for larger numbers of electrons [17].

Simply put, the electron molecule is formulated as follows. When electrons take a molecular configuration, we can think of the system as a molecule that executes vibrational motions (including quantum zero point motion) around the classical configuration on top of the rotational motion. This should be a good picture when the vibrational amplitude is not too large (as for large enough B).

Since we have a collection of fermions, we have in addition to impose a total antisymmetry in the wavefunction. So we can write the wavefunction as a rotational-vibrational state for N-electron system in a general form,

$$\Psi = \psi_{CM} \exp(-iL\chi) f_{J_{RM}, n_1...n_{2N-3}}(Q_1..., Q_{2N-3}) \psi_{\text{spin}}, \tag{1}$$

where ψ_{CM} is the centre-of-mass (CM) wavefunction, $\exp(-iL\chi)$ the wave function for rotation about the CM, f a vibrational wavefunction for the normal coordinates $\{Q_i\}$ and $n_1, ...n_{2N-3}$ are the numbers of quanta in each vibrational mode in the harmonic approximation, ψ_{spin} is the spin wavefunction.

Then, only for a specific combination of the quantum numbers the antisymmetrisation feasible. In particular, if the molecule has an m-fold symmetry, the vibrational wavefunction has to be an irreducible representation of C_m, and the angular momentum, L, has to take a specific (magic) series of values,

$$L \equiv 0(m/2) \,(\text{mod}\,m)\,\text{for}\,m\,\text{odd(even)},$$

when the vibrational state is in its ground state — only in that case does the permutation of fermions change the sign of wavefunction. This explains why we have a stabilisation energy when L takes the magic values, and why those magic values are dictated by the molecular symmetry.

When one considers the spin degrees of freedom, i.e., when we antisymmetrise the total wavefunction that include ψ_{spin}, the magic L values are linked with the total spin [12]. This is already apparent in the first numerical study of spin dependent correlation in quantum dots [18]. These molecules are characterized by a spin quantum number, k_s, where the spin wave function Ψ_{spin} is transformed to $\exp(-2\pi k_s i/m)\Psi_{\text{spin}}$ under the rotation of $2\pi/m$ for an m-fold symmetric molecule. Then the full criterion for the magic number reads

$$L + k_s \equiv 0\,(\text{mod}\,m)\,\text{for}\,m\,\text{odd}, \tag{2}$$
$$\equiv m/2\,(\text{mod}\,m)\,\text{for}\,m\,\text{even}.$$

The problem at hand, therefore, is how the magic S, which is linked with magic L, undergoes a series of jumps for each number, N, of electrons. If there are regions of B over which the magic values of S differ by more than $1/2$ between N- and $(N+1)$-electron systems, then we have a spin blockade. Since (i) how S jumps depends on the confining potential, and (ii) the electron molecule picture degrades for weaker B, we should have a numerical result. We have obtained the ground states (as well as low-lying excited states) with the exact diagonalisation. We consider a dot with a parabolic confining potential, where the motion of electrons is assumed to be completely two dimensional. The Hamiltonian is $\mathcal{H} = \mathcal{H}_s + \mathcal{H}_C$, where the single-electron part is

$$\mathcal{H}_s = \sum_n \sum_\ell \sum_\sigma \varepsilon_{n\ell\sigma} c^\dagger_{n\ell\sigma} c_{n\ell\sigma}, \tag{3}$$

$$\varepsilon_{n\ell\sigma} = (2n+1+|\ell|)\hbar \left(\frac{\omega_c^2}{4} + \omega_0^2\right)^{1/2} - \frac{1}{2}\ell\hbar\omega_c - g^*\mu_B B S_z, \tag{4}$$

where $\varepsilon_{n\ell\sigma}$ is the eigenenergy, $\hbar\omega_0$ represents the strength of the parabolic confinement potential, $\omega_c = eB/m^*c$ the cyclotron frequency, m^* the effective mass, μ_B the Bohr magneton, g^* the g-factor and S_z the z-component of the total spin. The interaction is

$$\mathcal{H}_C = \frac{1}{2} \sum_{n_1 \sim n_4} \sum_{\ell_1 \sim \ell_4} \sum_{\sigma_1 \sim \sigma_4} \langle n_1\ell_1\sigma_1, n_2\ell_2\sigma_2 | \frac{e^2}{\epsilon|\mathbf{r}_1 - \mathbf{r}_2|} | n_3\ell_3\sigma_3, n_4\ell_4\sigma_4\rangle$$
$$\times c^\dagger_{n_1\ell_1\sigma_1} c^\dagger_{n_2\ell_2\sigma_2} c_{n_4\ell_4\sigma_4} c_{n_3\ell_3\sigma_3}. \tag{5}$$

The Hamiltonian is written in second quantised form in the Fock-Darwin basis. As was shown by Fock and Darwin [19] earlier this century, when the confinement potential is parabolic, the one-electron states are just Landau's wavefunctions with the cyclotron radius modified into $\lambda = \sqrt{\hbar/(2m^*\Omega)}$, where $\Omega^2 \equiv \omega_0^2 + \omega_c^2/4$. Physically the states are localised on rings of width $\sim \lambda$ with a radius $R \simeq \sqrt{2(2n+|l|+1)}\lambda$. So R increases with angular momentum, while it decreases with magnetic field because λ approaches the magnetic length $\sqrt{\hbar c/eB}$ in the strong field limit.

2.3 Results and Discussion

We have numerically diagonalised the above Hamiltonian for four electrons in the confinement potential $\hbar\omega_0 = 2 \sim 6$ meV. This range is chosen to reproduce the addition energy spectrum in a dot realised in a mesa-etched, double-barrier InGaAs-AlGaAs heterostructure [21,20] In the numerical calculations we have used enough states (including higher Landau levels) in the basis to ensure convergence of the ground-state energy within 0.1%. We can see that the result (Fig.5) for the energy against L for a fixed value of B has a characteristic oscillation in the energy for each value of $S(= 0, 1$ or $2)$. This behaviour is surprisingly accurately reproduced by the electron-molecule picture as indicated in the Figure, where the agreement persists not only for strong B (large L) but down to moderate B (L).

When B is varied, the angular momentum L makes a series of jumps from a magic value to another, where the total spin S also jumps hand in hand with L. In the electron molecule picture we can see that L and S have to be related as

$S = 0$		$S = 2$
$k_s = 0$	$k_s = 2$	$k_s = 0$
(RVB^-)	(RVB^+)	
$L \equiv 2$	$L \equiv 0$	$L \equiv 2$

and the spin quantum number are indeed related through the magic-number criterion,

$$k_s = 0 \text{ for } L = 2 + 4 \times \text{integer}, \tag{6}$$
$$k_s = 2 \text{ for } L = 4 \times \text{integer}, \tag{7}$$

174   Hideo Aoki

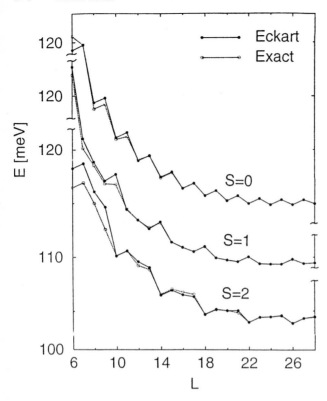

Fig. 5. The electron-molecule result for the magic-number plot for the total spin $S = 0, 1, 2$ for a single dot having four spinful electrons for $\hbar\omega = 4$ meV and $B = 20$ T[13]. Dashed lines, almost indistinguishable from the solid ones, represent the exact diagonalisation result.

for $N = 4$. We can make an intriguing identification, by looking at the spin density correlation function, that $(L, S) = (2, 0)$ is an RVB$^-$ state while $(4, 0)$ is an RVB$^+$, where the RVB's are defined, for a four-site cluster, as

$$\text{RVB}^-(k_s = 0): \quad \square - \square \quad = |{\uparrow\uparrow \atop \downarrow\downarrow}\rangle + |{\downarrow\downarrow \atop \uparrow\uparrow}\rangle + |{\uparrow\downarrow \atop \uparrow\downarrow}\rangle + |{\downarrow\uparrow \atop \downarrow\uparrow}\rangle$$

$$-2\left[|{\downarrow\uparrow \atop \uparrow\downarrow}\rangle + |{\uparrow\downarrow \atop \downarrow\uparrow}\rangle\right], \qquad (8)$$

$$\text{RVB}^+(k_s = 2): \quad \square + \square \equiv \boxtimes = |{\uparrow\uparrow \atop \downarrow\downarrow}\rangle + |{\downarrow\downarrow \atop \uparrow\uparrow}\rangle - |{\uparrow\downarrow \atop \uparrow\downarrow}\rangle - |{\downarrow\uparrow \atop \downarrow\uparrow}\rangle, \quad (9)$$

where $\square \equiv \frac{1}{\sqrt{2}}(|\uparrow\downarrow\rangle - |\downarrow\uparrow\rangle)$ is the spin-singlet pair. Namely, an RVB is defined as a linear combination of spin singlets with which the system (usually a molecule or a lattice) is covered, and here RVB$^+$ (RVB$^-$) has two singlets combined with a $+(-)$ sign. In other words RVB$^+$ from RVB$^-$ is that the former lacks the Néel components (the last two terms in RVB$^-$) and has the extra phase factor -1 for a $2\pi/4$ rotation. Although what we have here is a system in a continuous space, totally different from lattice fermions such as the Hubbard model for which RVB is usually conceived, the electron-molecule formation has brought about such spin configurations. The total angular momentum L, and the total spin $S(N)$ of

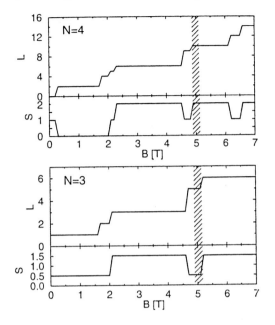

Fig. 6. The total angular momentum and total spin of the ground state for a single quantum dot with $N = 3$ (lower panel) or $N = 4$ electrons (upper)[4]. The confinement potential is assumed to be parabolic with $\hbar\omega_0 = 3.0$ meV. The spin-blockade region is indicated by shading.

the ground state for three- and four-electron systems are plotted in Fig. 6, and we can see how the magic L values go hand in hand with S: As the magnetic field increase, the ground state evolves as $L(S) = 1 \rightarrow 2 \rightarrow 3(1/2 \rightarrow 1/2 \rightarrow 3/2)$ for $N = 3$, $L(S) = 0 \rightarrow 2 \rightarrow 3 \rightarrow 4 \rightarrow 5 \rightarrow 6(1 \rightarrow 0 \rightarrow 1 \rightarrow 0 \rightarrow 1 \rightarrow 2)$ for $N = 4$.

If we then plot the difference in the total spin, $S(4) - S(3)$, against the magnetic field in the bottom panel of Fig. 7, the spin blockade condition,

$$|S(N+1) - S(N)| > \frac{1}{2}, \tag{10}$$

is indeed fulfilled: S jumps from $3/2$ to 0 in the region 4.96 T$< B <$ 5.18 T.

In the spin-blocked region, the conduction, blocked at zero temperature, has to occur through an $S = 1$ excited state for $N = 4$ at finite temperatures. If the excited states are well separated in energy (≥ 0.1 meV, typical experimental resolution) from the ground state for both of the $(N - 1)$- electron and N-electron states, the spin blockade should be observed in the Coulomb diamond, which is the differential conductance plotted in the plane of source-drain voltage and gate voltage. To check this we have calculated the three lowest excitation energies and their total spins for the $N = 4$ quantum dot in Fig. 7. The lowest excited states for $N = 3$ and for $N = 4$ both lie about 0.06 meV above the ground state around $B = 5.1$ T in the spin-blockade region. We can make this separation larger (~ 0.1 meV) for stronger confinement potentials (e.g., 0.09 meV around $B = 7.4$ T for $\hbar\omega_0 = 8.0$ meV). Such confinement potentials may be realised in a gated vertical quantum dot [21].

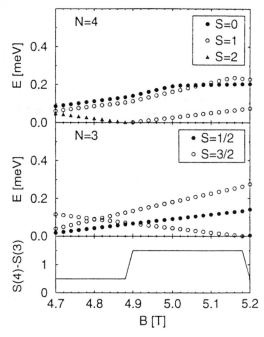

Fig. 7. Top: Excitation energies for $N = 4$ single dot. Middle: The same for $N = 3$. Bottom: Difference in the total spin, $S(4) - S(3)$, between the ground state for $N = 3, 4$[4].

The link between the magic L and total S and subsequent spin blockade appears for other numbers of electrons as well, e.g., between $(L, S) = (2, 0)$ state for $N = 2$ and $(6, 3/2)$ state for $N = 3$ for 14.1 T $< B < 14.8$ T.

One of the consequences of permutational symmetry in the electron molecule is that there are actually $N!$ equilibrium configurations, each corresponding to a different permutation of the N electrons. Some of these configurations are connected by rotations but others are not. For example, triangular configurations of 3 electrons, $\frac{1}{23}$ and $\frac{3}{12}$, can be rotated into each other but not into $\frac{2}{13}$. So the configurations that cannot be rotated into each other coexist as degenerate states. When the 'atoms' (Landau's orbits in the present case) begin to overlap, quantum mechanical tunnelling will break the degeneracy, which should become significant for weaker B, hence for smaller angular momenta. So the electron molecule picture breaks down there, seen in Fig.5 as a deviation from Eqn(3). [22]

We can extend [23–26] the above line of approach to double dots, where two dots are stacked in the vertical direction as have been fabricated [27] in semiconductor heterostructures. Due to an added degree of freedom, we can have steric (three-dimensional) electron molecules, such as a tetrahedral molecule for a four-electron system. This occurs depending on the relative strengths of intra-layer correlation and inter-layer correlation and the inter-layer electron tunnelling, which are controlled by the inter-layer separation. The spin state again dominates the way in which the correlation effect is exerted, so we have magic numbers specific to the double-dot systems and resulting low-to-high spin transitions [23]. One appearance of this is the spin blockade condition becomes less stringent, due to the flexibility introduced by the layer degrees of freedom.

3   Tomonaga–Luttinger System in Magnetic Fields

3.1   Introduction

1D is the lowest spatial dimension in which we have a fermion system with a
Fermi energy and a Fermi wavenumber. In general the effect of the electron-
electron interaction is the stronger the lower the dimension due to the strong
constraint in the phase space. In 1D the interaction is in fact so crucial that the
system becomes universally what is called the Tomonaga-Luttinger (TL) liquid,
a time-renowned many-body theory half a century old, as far as low-lying excitations are concerned no matter how small the interaction may be [28]. A most
striking feature is the spin-charge separation, which also dominates the transport properties [29] in the following sense. The low-temperature conductivity of
the dirty TL liquid as studied by Luther and Peschel [30] exhibits a power-law
temperature dependence, $\sigma(T) \sim T^{2-K_\rho-K_\sigma}$, where K_ρ and K_σ are the critical
exponents related to the charge and spin, respectively, vital parameters in the
TL theory. The power law comes from the Fermi surface degraded from a jump
into a power-law singularity in the TL liquid, while the critical exponents (which
are functions of the interaction) enter as a sum of K_ρ for the phase representing
the charge degrees of freedom and K_σ for the spin phase (which is actually unity
for spin-independent interactions). A pioneering experiment [31] for high-quality
quantum wires supports this result.

Now we can raise an intriguing question: what happens if we degrade the spin-
rotational (SU(2)) symmetry in some way? This is indeed realised by applying
a magnetic field, which makes the Fermi velocity different for different spin
directions with the Zeeman splitting (inset, Fig.9).

In other words, we introduce here a Tomonaga-Luttinger model comprising
two components with different Fermi velocities.

In this section we shall show that the degraded spin-charge separation will
make even the quantities such as the density-density correlation involve spins,
which can manifest itself in the transport. Namely, the ratio of the up-spin
and down-spin conductivities in a dirty system can become very large at low
temperatures, leading such a drastic effect as spin-polarised currents [5].

The generation of spin-polarised currents itself has been of a long-standing interest for academic [32] as well as practical points of view, where typical applications include spin-polarised STM [33] or the Mott detector [34]. Fasol and Sakaki
[35] have suggested that in the spin-orbit split bands of GaAs quantum wires the
curvature in the band dispersion (as opposed to the linearised dispersion in the
TL model) will make the relaxation time due to the electron-electron interaction
spin-dependent and consequently make the outgoing current spin-polarised. The
mechanism proposed here is by contrast purely electron-correlation originated,
where the ratio, $\sigma_\uparrow/\sigma_\downarrow$, even diverges as $T \to 0$ when localisation is ignored.

3.2   Formulation

We first consider a clean, two-band Tomonaga-Luttinger model in a magnetic
field. We start with the formulation developed for the equilibrium case [36] or

the Fermi-edge singularity problem [37] to consider transport properties. The Hamiltonian is given by

$$\mathcal{H}_{\text{clean}} = \mathcal{H}_0 + \mathcal{H}_{\text{int}}, \tag{11}$$

with the non-interacting Hamiltonian being

$$\mathcal{H}_0 = \sum_{k,s,i} v_{Fs}[(-)^{i+1}k - k_{Fs}]c_{iks}^\dagger c_{iks}, \tag{12}$$

where $c_{iks}^\dagger$ creates a right-going ($i = 1$) or left-going ($i = 2$) electron with momentum k and spin s, and L is the system size. Here $v_{F\uparrow}$ ($v_{F\downarrow}$) is the Fermi velocity of the up (down) spin in a magnetic field B, and we denote the average $v_0 \equiv (v_{F\uparrow} + v_{F\downarrow})/2$ and the difference $\Delta v \equiv (v_{F\uparrow} - v_{F\downarrow})/2$. Similarly, $k_{F\uparrow}$ ($k_{F\downarrow}$) is the Fermi wavenumber of the up (down) spin.

The interaction Hamiltonian is given by

$$\mathcal{H}_{\text{clean}} = \frac{v_\rho}{4\pi} \int dx \left\{ \frac{1}{K_\rho}[\partial_x\theta_+(x)]^2 + K_\rho[\partial_x\theta_-(x)]^2 \right\}$$
$$+ \frac{v_\sigma}{4\pi} \int dx \left\{ \frac{1}{K_\sigma}[\partial_x\phi_+(x)]^2 + K_\sigma[\partial_x\phi_-(x)]^2 \right\}$$
$$+ \frac{\Delta v}{2\pi} \int dx \left\{ [\partial_x\theta_+(x)][\partial_x\phi_+(x)] + [\partial_x\theta_-(x)][\partial_x\phi_-(x)] \right\}, \tag{13}$$

where $\theta_+(x)$ ($\phi_+(x)$) is the phase field for the charge (spin) degrees of freedom, while $\theta_-(x)$ ($\phi_-(x)$) are their dual.

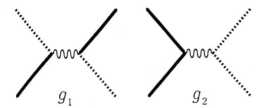

Fig. 8. Feynmann diagrams representing the backward (g_1) and forward (g_2) scattering processes. Solid (dotted) lines stand for the left(right)-going branch.

The spin-charge separated part (first two lines) is the usual phase Hamiltonian, while the charge (θ) and spin (ϕ) degrees of freedom are indeed coupled (the last line) for $\Delta v \neq 0$, in a sharp contrast to the usual TL Hamiltonian. Here K_ρ, the critical exponent for the charge, is given by

$$K_\rho = \sqrt{\frac{1 - \tilde{g}_\rho}{1 + \tilde{g}_\rho}}, \tag{14}$$

where $\tilde{g}_\rho \equiv g_\rho/2\pi v_F, g_\rho = 2g_2 - g_1$ with $g_2(g_1)$ being forward (backward) scattering coupling constants (Fig 8; g's are U in the Hubbard model for small U). The critical exponent of the spin phase, $K_\sigma = \sqrt{(1 - \tilde{g}_\sigma)/(1 + \tilde{g}_\sigma)}$ ($g_\sigma = g_1$), is usually unity, since g_1 is renormalised into zero for spin-independent electron-electron interactions unless the system has a spin gap for some specific reason (as in ladders). $v_\rho(v_\sigma)$ is the charge (spin) velocity with $v_\nu = v_F\sqrt{1 - \tilde{g}_\nu^{\;2}}$. We assume that the coupling constants between electrons have no significant magnetic-field dependences. We have also neglected the Umklapp scattering processes, since they have large momentum transfers.

We can diagonalise $\mathcal{H}_{\text{clean}}$, as is done for the electron-hole system in a two-channel Tomonaga-Luttinger study of the excitonic phase [38], with a linear transformation to two new phase fields,

$$\begin{pmatrix} \theta_+(x) \\ \phi_+(x) \end{pmatrix} = \begin{pmatrix} \cos\alpha & -\frac{1}{y}\sin\alpha \\ y\sin\alpha & \cos\alpha \end{pmatrix} \begin{pmatrix} \tilde{\theta}_+(x) \\ \tilde{\phi}_+(x) \end{pmatrix}, \tag{15}$$

where α is the 'rotation angle in the spin-charge space' ($\propto \Delta v$ for small Δv) with $\tan 2\alpha = 2(\Delta v/v_0)\sqrt{2(K_\rho^{-2} + 1)/(K_\rho^{-2} - 1)}$ and $y^2 = \frac{1}{2}(K_\rho^{-2} + 1)$. The diagonalised phases have gapless, linear dispersions with velocities that depend on α.

Now we can turn to the calculation of the conductivity in a dirty system. We then add to the Hamiltonian the impurity scattering,

$$\mathcal{H}_{\text{imp}} = \sum_s \sum_l \int dx \, \rho_s(x) \, u(x - x_l), \tag{16}$$

where $u(x - x_l)$ is the impurity potential situated at x_l and $\rho_s(x)$ is the density operator of spin s electrons, whose phase representation is $\rho_s = \frac{1}{2\pi}\partial_x(\theta_+ + s\phi) + \frac{1}{\pi\Lambda}\cos(2k_{Fs}x + \theta_+ + s\phi)$ with Λ being a short-range cutoff. The conductivity, σ_s, for spin s is given by $\sigma_s = n_e e^2 \tau_s/2m_s^*$, where τ_s is the relaxation time for spin s, n_e the density of electrons, and $m_s^* \propto v_s^{-1}$ the effective mass of the spin s subband. In 1D we have $n_e = 2k_F/\pi$, where we ignore the trivial magnetic-field dependence of $k_{F\uparrow}$ and $k_{F\downarrow}$ to single out the effect of $v_{F\uparrow}/v_{F\downarrow} \neq 1$.

We can calculate the relaxation time in the Mori formalism for the conductivity [39,40] in the second order in $\mathcal{H}_{\text{imp}}$ as

$$\frac{1}{\tau_s} \approx 4\pi v_{Fs} n_i u^2 (2k_{Fs}) \sum_q \lim_{\omega \to 0} \frac{\text{Im} \Pi_s(2k_{Fs} + q, \omega)}{\omega}, \tag{17}$$

where n_i is the density of impurities and $u(q)$ the Fourier transform of $u(x)$. Here Π_s is the density-density correlation function for spin s, which is related to the density operator $\rho_s(x)$ via

$$\lim_{\omega \to 0} \sum_q \frac{\text{Im} \Pi_s(2k_{Fs} + q)}{\omega} = \frac{1}{2T} \sum_{s'} \int_{-\infty}^{\infty} dt \, \langle \rho_s(0, t)\rho_{s'}(0, 0)\rangle. \tag{18}$$

In the summation over the spin s' we can readily show that the cross term, $\langle \rho_\uparrow(0, t)\rho_\downarrow(0, 0)\rangle$, vanishes. Then the conductivity becomes a sum of the two spin components,

$$\sigma_s(T) = \sigma_0 \left(\frac{v_{Fs}}{v_0}\right)^2 \left(\frac{T}{\omega_F}\right)^{2-K_s}, \qquad (19)$$

each of which has a simple power-law temperature dependence as in the usual TL theory. Here $\sigma_0 \equiv \sigma(T = \omega_F)$ (whose dependence on k_{Fs} is again ignored here) and $\omega_F \sim \varepsilon_F$ is the high-energy cutoff. [41]

The exponent K_s, which now depends on the spin direction, is given by

$$K_s = (\cos\alpha + sy\sin\alpha)^2 \tilde{K}_\rho + \left(\cos\alpha - \frac{s}{y}\sin\alpha\right)^2 \tilde{K}_\sigma, \qquad (20)$$

This formula involves both the charge exponent, $\tilde{K}_\rho$, for the phase $\tilde{\theta}$ and the spin exponent, $\tilde{K}_\sigma$, for $\tilde{\phi}$ as given by

$$\tilde{K}^2_{\rho(\sigma)} = y^{\mp 2} \frac{K_\rho^{-2} + 3 \pm (K_\rho^{-2} - 1)\sqrt{1 + \tan^2 2\alpha}}{3K_\rho^{-2} + 1 \pm (K_\rho^{-2} - 1)\sqrt{1 + \tan^2 2\alpha}} \qquad (21)$$

where the upper (lower) sign corresponds to $\tilde{K}_\rho$ ($\tilde{K}_\sigma$).

Thus the electron-electron interaction does indeed make the conductivity dependent on the spin. Since the conductivity for each spin has a power-law in T with a spin-dependent power, the spin dependence becomes more enhanced at lower temperatures with the ratio, $\sigma_\uparrow/\sigma_\downarrow \propto T^{-(K_\uparrow - K_\downarrow)}$, diverging for $T \to 0$, which implies a spin-polarised current.

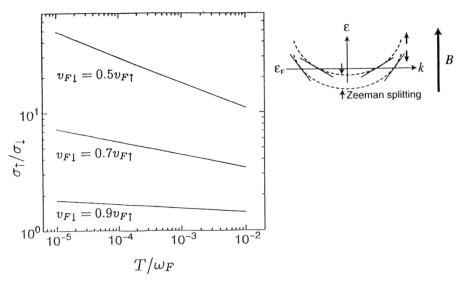

Fig. 9. The result for the temperature dependence of the ratio $\sigma_\uparrow/\sigma_\downarrow$ with a fixed interaction $\tilde{g}_\rho = 0.5$, and with several values of $v_{F\uparrow}/v_{F\downarrow}$[5]. The inset depicts the dispersion around E_F which is considered in the Tomonaga-Luttinger theory.

3.3   Results and Discussion

We display in Fig.9 the temperature dependence of $\sigma_\uparrow/\sigma_\downarrow$ numerically calculated for various values of $v_{F\uparrow}/v_{F\downarrow}$ for a fixed electron-electron interaction g.

Figure 10 shows the dependence of $\sigma_\uparrow$ ($\sigma_\downarrow$) on the ratio $v_{F\uparrow}/v_{F\downarrow}$ at a fixed temperature and a fixed g.

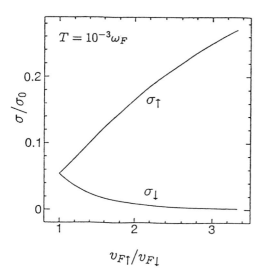

Fig. 10. The result for the dependence on $v_{F\uparrow}/v_{F\downarrow}$ (keeping $v_{F\uparrow} + v_{F\downarrow}$=const.) of the conductivity $\sigma_\uparrow$ and $\sigma_\downarrow$ with a fixed temperature $T = 10^{-3}\omega_F \sim 100$ mK and with a fixed interaction $\tilde{g}_\rho = 0.5$.[5]

These results show that the more conductive channel is the spin having a larger v_F, since $K_\uparrow > K_\downarrow$ for $\Delta v \propto (v_{F\uparrow} - v_{F\downarrow}) > 0$. Physically, the effective electron-electron interaction is smaller (larger) in the lighter (heavier) spin sub-band, since it is $\tilde{g} \equiv g/2\pi v_F$, the ratio of the electron-electron interaction to v_F, that enters in the TL theory. In this sense the present result is consistent with the property in a single-component TL liquid that the electron-electron repulsive interaction suppresses the conductivity [30].

We can further interpret the result, if we map the single TL model for spin 1/2 electrons onto a double-wire system of spinless electrons by regarding the spin up(down) sector the wire 1(2). The present situation is then a generalisation of our previous model [40], where we have considered two wires (chains) having intrachain and interchain interactions in the absence of interchain tunnelling. When we generalise this work by making the two chains inequivalent with different v_F for two chains, we can regard the system as consisting of a wire with a strong intra-chain coupling constant, g/v_F, a weak intra-chain coupling, and an inter-chain interaction. This translates into the interactions between the up-spins, down-spins and the up-down interaction, respectively, where all the interactions are the same for a usual SU(2) symmetric interaction (such as the Hubbard U) but become different in the presence of a Zeeman splitting. The present result that the spin sector that has a smaller g/v_F (i.e., a weaker repulsion) is more conductive is consistent with the double-chain result that the intra-chain repulsion is the rate-determining (conductivity-suppressing) process in the charge-charge correlation function, while the effect of the inter-chain interaction, which incidentally enhances the conductivity, is only of the second order.

Finally let us make a comment on the Anderson localisation. At very low temperatures,

$$k_B T < \hbar v_F / \lambda \, (\lambda : \text{mean free path}),$$

we must treat the impurity scattering beyond the simple perturbation to consider the renormalisation effect. We have studied [42] such a renormalisation following the method given in Ref. [43]. The renormalisation flow is divided into

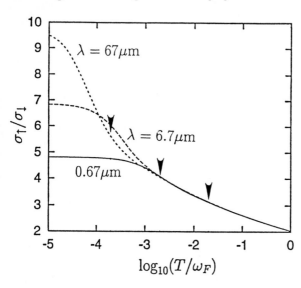

Fig. 11. The temperature dependence of the ratio $\sigma_\uparrow / \sigma_\downarrow$ with fixed $v_{F\uparrow} / v_{F\downarrow} = 0.7$ and $\tilde{g}_\rho = 0.5$ for various values of the mean free path $0.67\mu\text{m} \le \ell \le 67\mu\text{m}$[42], where $\ell = 67\mu\text{m}$ is taken from an estimate for a 2D system in Ref.[31]. Arrows indicate the position of $k_B T = \hbar v_F / \lambda$ for respective value of λ.

three regimes. For the usual case of repulsive interactions (or weak attractive) interactions with $K_\uparrow, K_\downarrow < 3$, the impurity scatterings for both spins are monotonically enhanced at low temperatures, so that both spin bands are, as expected in 1D, Anderson-localised. If the attraction is strong enough, $K_\uparrow > 3$, at least one spin band survives localisation. For repulsive interactions, the critical exponents K_ρ and K_σ will then be renormalised to zero for $T \to 0$, so that the behaviour of $\sigma_s(T)$ will change from T^{2-K_s} to T^2. Accordingly, the ratio $\sigma_\uparrow / \sigma_\downarrow$ (Fig.11) for $T \to 0$ saturates from $T^{-(K_\uparrow - K_\downarrow)}$ to a constant below a crossover temperature $k_B T \sim \hbar v_F / \lambda$. Thus the ratio, although no longer divergent, can still be significantly large.

We believe these novel many-body effects can be experimentally measured in quantum wires by taking low enough band filling (which will make v_F different enough between up- and down-spin) in a strong enough magnetic field. In the case of the usual electron-doped GaAs quantum wires the g-factor ($\simeq -0.4$ in the bulk) is unfortunately too small to attain a sufficient Zeeman splitting. However, if we can prepare a material e.g. InSb whose g-factor is as large as $\simeq -50$ in the bulk [44], the Zeeman splitting in a typical magnetic field of 1T amounts to

$g^* \mu_B B \simeq 3.0$ meV. In such a case a significant deviation of $v_\uparrow / v_\downarrow$ from unity may be expected in quantum wires.

4 Negative Magnetoresistance Originating from Electron Correlation

4.1 Introduction

Now we come to a totally different appearance of the correlation effect in which spin degrees of freedom appear in a transport property. Namely, we propose, inspired from an experimentally examined organic conductor, a novel mechanism for negative magnetoresistance for repulsively interacting electrons on a lattice whose band dispersion contains a *flat* portion (a flat bottom below a dispersive part here) [6]. This mechanism is applicable to *any* spatial dimension, but we exemplify it in 2D here.

In general, negative magnetoresistance (i.e., the system becoming more conductive in the presence of B) has provided fascination in diverse classes of systems, such as the impurity band in semiconductors [45] or, most recently, Mn oxides [46]. The problem is the interplay of the transport and the spin structure, and how the spins and carriers respond to magnetic fields can vary from system to system. The problem posed here is, since the spin is quite generally a key ingredient in the physics of correlated electron systems, can we use this for the negative magnetoresistance. In particular we start from an observation that the itinerant (metallic) ferromagnetism has been a central problem from the days of Kanamori [47], Hubbard [48] and Gutzwiller. [49] It is becoming increasingly clear that the ferromagnetism is rather difficult to realise as one takes account of correlation effects beyond mean-field approximations, but a large density of states at the Fermi level (which is indeed the situation originally considered by Kanamori [47]) does tend to favour ferromagnetism.

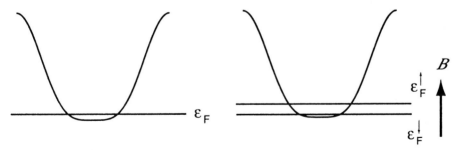

Fig. 12. Basic idea for the negative magnetoresistance in a system with a partially flat dispersion for $B = 0$ (left panel) and $B \neq 0$ (right).

The basic idea is the following (Fig.12). We consider repulsively interacting electrons on a band, whose one-electron dispersion has a flat part in an otherwise dispersive band. When the Fermi level lies in the flat part, the electron

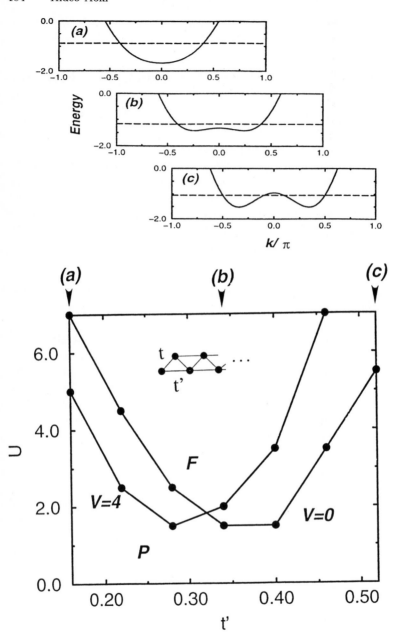

Fig. 13. Density matrix renormalization group (DMRG) result for the phase diagram (P: paramagnetic, F: ferromagnetic) against U and t' for the extended Hubbard model on a 1D t-t' lattice (inset) for $n = 0.4$ when the off-site repulsion V is varied from 0 to 4.[54] One-electron band structures are displayed in the accompanying panels for typical values of t' labeled with (a),(b),(c), where the horizontal dashed lines indicate the Fermi level.

correlation should cause ferromagnetic spin fluctuations to develop. The situation of the large density of states $D(E_F)$ around the band bottom has received a renewed interest recently, and tendencies towards ferromagnetism have been confirmed for 2D (square lattice with an appreciable second-nearest neighbour hopping, etc) and 3D (fcc, etc) systems [50–52] as well as in 1D (chain with an appreciable second-nearest neighbour hopping [53–55] or a model atomic quantum wire [56]). We show an example of the phase diagram [54] obtained for the 1D system in Fig.13, where we do have a ferromagnetic phase for strong repulsive interactions, U, when the band bottom is flat enough.

The spin susceptibility is then enhanced due to ferromagnetic fluctuations, so that a weak magnetic field is enough to drive the system into a significant magnetisation (spin polarisation). The E_F for the majority spins will then be pushed into the dispersive (i.e., lighter-mass) part of the band, and we expect that the system becomes more conductive, resulting in a negative magnetoresistance. Thus the mechanism for negative magnetoresistance proposed here combines an effect of the electron correlation (proximity to ferromagnetism) and the band structure (coexistence of flat and dispersive parts). We shall then analytically confirm the idea by calculating the conductivity in magnetic fields for the repulsive Hubbard model on a typical tight-binding model having a partly flat band.

This idea has been conceived from a recent experimental result on a certain class of organic conductors, called τ-type conductors, for which Murata et al [57] have observed negative mangetoresistance, see below. The band structure calculation indicates that the dispersion is indeed flat at the bottom along certain directions.

4.2   Formulation

We formulate our idea for the single-band Hubbard model, a simplest model for repulsive electron correlation, on a simplest flat-bottomed tight-binding model. Namely, when the square lattice have a large second nearest neighbour hopping, $t' \sim |t|$, on top of t between nearest neighbours, the one-electron dispersion becomes

$$\varepsilon_{\mathbf{k}}^0 = 2t \left(\cos k_x + \cos k_y \right) + 4t' \cos k_x \cos k_y, \tag{22}$$

which is flat along $k_x = 0$ and $k_y = 0$ (right panel of Fig. 14) for $t' \simeq 0.5|t|$ with the van Hove singularity pushed down to the bottom.

For this lattice we consider the Hubbard Hamiltonian,

$$\mathcal{H} = \sum_{i,j,\sigma} t_{ij} c_{i\sigma}^\dagger c_{j\sigma} + U \sum_i n_{i\uparrow} n_{i\downarrow} \tag{23}$$

in standard notations.

For $t' \simeq 0.5$ (we take $|t| = 1$ hereafter) Hlubina has shown, with the T-matrix approximation for the Hubbard model, that the ground state is fully spin-polarised for the band filling $n \sim 0.4$. [50,51] Since the T-matrix approximation

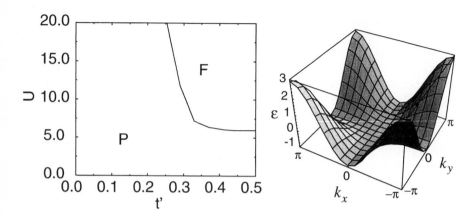

Fig. 14. The phase diagram (P: paramagnetic, F: ferromagnetic) for the 2D t-t' Hubbard model, determined by the exact diagonalisation of 8 electrons in 4 × 4 system.[6] The right panel is the one-electron dispersion for $t = -1, t' = 0.5$.

is valid only in the low enough densities and for small Hubbard repulsion U, here we start with obtaining the diagram for large U with the exact diagonalisation of a finite system (4×4 sites with 8 electrons with the band quarter-filled).

Identifying ferromagnetism from total $\mathbf{S}^2$ in finite systems requires in general some care as we have revealed in previous publications [55]. Namely, ferromagnetic states in an itinerant electron model can accompany a *spiral spin state* in finite systems. The spiral state is a spin singlet state having the spin correlation length as large as the linear dimension of the system, which can lie in energy below the ferromagnetic state. We found for the present model that the ground state of the 16-site system has always $\mathbf{S}^2 = 0$ in periodic boundary conditions in both x and y, but if we look at the spin structure $S(\mathbf{q})$ (Fourier transform of $\langle \mathbf{S}_i \cdot \mathbf{S}_j \rangle$) the state turns out to be indeed spiral (i.e., $S(\mathbf{q})$ peaked at $(0, \pm\pi/L)$ and $(\pm\pi/L, 0)$) for large U and $t' \simeq 0.5$.

While we can identify the spiral state as ferromagnetic in the thermodynamic limit, it has been shown [58] that a faster approach to the thermodynamic limit is attained if we adopt an appropriate boundary condition (periodic in one direction and antiperiodic in another, which makes the ground state of the noninteracting system an open shell) to selectively push the spiral state above the ferromagnetic state in energy. If we look at the phase diagram thus obtained against t' and U in Fig. 14, the fully polarised ferromagnetic region is seen to exist for $t' \sim 0.5$ and $U > 6$.

Keeping this phase diagram in mind, we now obtain the conductivity in a magnetic field B. The conductivity of the interacting system is calculated here with the fluctuation exchange approximation (FLEX). The FLEX, introduced by Bickers *et al.* [59], treats spin and charge fluctuations (Fig.15) by starting from a set of skeleton diagrams for the Luttinger-Ward functional, based on the idea

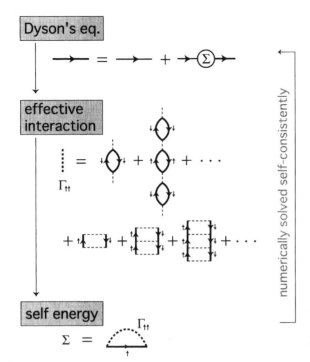

Dyson's eq.

effective interaction

$\Gamma_{\uparrow\uparrow}$

self energy

$\Sigma =$

numerically solved self-consistently

Fig. 15. The Feynmann diagrams that are taken into account in the FLEX approach.

of Baym and Kadanoff [60]. A (k-dependent) self energy is computed from RPA-type bubble and ladder diagrams self-consistently. A recent example of successful applications of FLEX is a reproduction by Kontani *et al* [61] of the anomalous behaviours in the resistivity and the Hall coefficient of High-T_C cuprates that are modelled by the 2D Hubbard model with strong antiferromagnetic fluctuations.

The dc conductivity is given in the Kubo formula as

$$\sigma_{\mu\nu} = \lim_{\omega \to 0} e^2 \sum_{\sigma\sigma'} \int \frac{d\mathbf{k}d\mathbf{k}'}{(2\pi)^6} v^0_{\mathbf{k}\mu} v^0_{\mathbf{k}'\nu} \frac{\mathrm{Im} K_{\mathbf{k}\mathbf{k}'\sigma\sigma'}(\omega + i\delta)}{\omega},$$

where $K_{\mathbf{k}\mathbf{k}'\sigma\sigma'}$ is the Fourier component of the retarded two-particle Green's function and $v^0_{\mathbf{k}\mu} = \partial \varepsilon^0_{\mathbf{k}}/\partial k_\mu$ is the unperturbed velocity. We set $\hbar = 1, k_B = 1$ hereafter. If we follow Eliashberg [62] for the analytic continuation of $K(\omega + i\delta)$ from $K(i\omega_n)$, where ω_n is the Matsubara frequency, the conductivity per spin reads, for long enough life time of the quasi-particle,

$$\sigma_{xx} = e^2 \int \frac{d\mathbf{k}}{(2\pi)^3} \left(-\frac{\partial f}{\partial \varepsilon} \right) \frac{v_{\mathbf{k}x} J_{\mathbf{k}x}}{2\gamma_{\mathbf{k}}}, \qquad (24)$$

where $v_{\mathbf{k}x}$ is the dressed velocity, $J_{\mathbf{k}x}$ the current, $\gamma_{\mathbf{k}}$ the damping constant of the quasiparticle, and f the Fermi distribution function.

Note that the conductivity of a clean system of interacting electrons can be finite even though the interaction is an internal force, since the momentum is dissipated through Umklapp processes. Conversely, the conductivity should diverge

when the Umklapp processes are turned off, and Yamada and Yosida [63] have
shown that this is guaranteed only if we consider the vertex correction for the
current $J_{\mathbf{k}}$ appropriately, so the correction should be included for a consistent
treatment. In the FLEX, which is an approximation conserving the total number of electrons and momentum, three types of diagrams are taken into account
for irreducible vertex [61], namely two Aslamasov-Larkin (AL) terms and one
Maki-Thompson (MT) terms. Kontani et al [61] have shown that, when antiferromagnetic fluctuations are dominant, the AL contribution can be neglected.
We can extend this argument to show that the AL term is also negligible when
ferromagnetic fluctuations are dominant. Hence we consider only the MT term.

Let us now discuss the conductivity in magnetic fields. We have then to add
the Zeeman term, $h \sum_{\mathbf{k}\sigma} \mathrm{sgn}(\sigma) c_{\mathbf{k}\sigma}^{\dagger} c_{\mathbf{k}\sigma}$, to the Hamiltonian, where $h \equiv g^* \mu_B B$
with $g^* \simeq 2$. Green's function and other quantities then become σ-dependent.
To concentrate on the effect of the Zeeman splitting we assume here that the
direction of the magnetic field is parallel to the current, so that we do not have
to take account of the effect of the field on orbital motions.

The conductivity is obtained from the Bethe-Salpeter equation, which is
a simple extension of the equations derived by Kontani et al for the spin-
independent case [61] to the present spin-dependent case. If we go back to a
more general form than Eq.(24) since it is not obvious whether quasiparticles
are well-defined when $D(E_F)$ is large, the diagonal conductivity reads

$$\sigma_{xx} = \qquad\qquad\qquad\qquad\qquad\qquad\qquad\qquad\qquad\qquad\qquad (25)$$

$$= e^2 \sum_{\mathbf{k},\sigma} \int \frac{d\varepsilon}{\pi N} \left(-\frac{\partial f(\varepsilon)}{\partial \varepsilon} \right) \left\{ |G_{\mathbf{k}\sigma}(\varepsilon)|^2 v_{\mathbf{k}x\sigma} J_{\mathbf{k}x\sigma}(\varepsilon) - \mathrm{Re} \left[G_{\mathbf{k}\sigma}^2(\varepsilon) v_{\mathbf{k}\sigma}^2(\varepsilon) \right] \right\},$$

$$J_{\mathbf{k}x\sigma}(\omega) = v_{\mathbf{k}x\sigma}(\omega) + \sum_{\mathbf{q}\sigma'} \int \frac{d\varepsilon}{2\pi N} \left[\coth \frac{\varepsilon - \omega}{2T} - \tanh \frac{\varepsilon}{2T} \right]$$

$$\times \mathrm{Im} V_{\mathbf{k}-\mathbf{q},\sigma\sigma'}(\varepsilon - \omega + i\delta) |G_{\mathbf{q}\sigma'}(\varepsilon)|^2 J_{\mathbf{q}x\sigma'}(\varepsilon),$$

where N is the number of sites, $G_{\mathbf{k}\sigma}(\omega)$ the dressed Green's function and the
velocity $v_{\mathbf{k}x\sigma} = (\partial/\partial k_x)[\varepsilon_{\mathbf{k}}^0 + \mathrm{Re}\Sigma_{\mathbf{k}\sigma}(\omega = 0)]$ with $\Sigma_{\mathbf{k}\sigma}(\omega)$ being the self energy.
The second term for σ_{xx} (Eqn.(25)), the 'incoherent term' as derived by Kontani
et al, contributes when the quasi-particle picture is invalid (i.e., $\gamma_{\mathbf{k}} \sim T$).

The kernel, $V_{\mathbf{k}\sigma\sigma'}(\omega)$, which contains the effect of fluctuation exchanges, can
be obtained by an analytic continuation of $V_{\mathbf{k}\sigma\sigma'}(i\omega_n)$ [64],

$$V_{\mathbf{k}\uparrow\uparrow}(i\omega_n) = \frac{U^2 \chi_{\mathbf{k}\uparrow\uparrow}^0(i\omega_n)}{1 - U^2 \chi_{\mathbf{k}\uparrow\uparrow}^0(i\omega_n) \chi_{\mathbf{k}\downarrow\downarrow}^0(i\omega_n)} - \frac{U^2}{2} \chi_{\mathbf{k}\uparrow\uparrow}^0(i\omega_n),$$

$$V_{\mathbf{k}\uparrow\downarrow}(i\omega_n) = \frac{U^2 \chi_{\mathbf{k}\uparrow\downarrow}^0(i\omega_n)}{1 - U \chi_{\mathbf{k}\uparrow\downarrow}^0(i\omega_n)} - \frac{U^2}{2} \chi_{\mathbf{k}\uparrow\downarrow}^0(i\omega_n) + U,$$

where $\chi_{\mathbf{k}\sigma\sigma'}^0 = -(T/N) \sum_k G_\sigma(k+q) G_{\sigma'}(k)$ with $k \equiv (\mathbf{k}, i\omega_n)$.

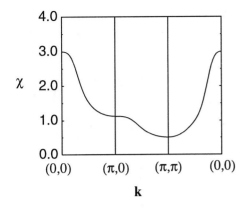

Fig. 16. The wave number dependence of $\chi_\mathbf{k}^{\text{RPA}}$ for $t' = 0.5, U = 2, n = 0.4, T = 0.1$.[6]

4.3   Results and Discussion

Let us now present the results for the case of $t' = 0.5, U = 2$ with the band filling $n = 0.4$, which falls upon the paramagnetic region close to the ferromagnetic boundary in the phase diagram, Fig.14. We first check that we do have strong ferromagnetic fluctuations. Figure 16 shows the wave number dependence of the spin susceptibility, $\chi_\mathbf{k}^{\text{RPA}} = 2\chi_\mathbf{k}^0(\omega = 0)/(1 - U\chi_\mathbf{k}^0(\omega = 0))$ for $T = 0.1$. We can see that there is indeed a peak around Γ $(\mathbf{k} = \mathbf{0})$.

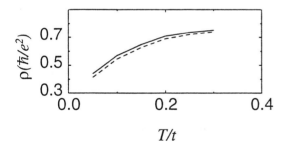

Fig. 17. The temperature dependence of the resistivity of the t-t' Hubbard model with the vertex corrections for $t' = 0.5, U = 2, n = 0.4$ in zero magnetic field ($h = 0$, full line) and in a magnetic field ($h = 0.05$, dashed line).[6]

To confirm that we are sitting close to the ferromagnetic boundary (i.e., the susceptibility enhanced), we have computed the static, uniform magnetic susceptibility also from the derivative $\chi \equiv \partial\langle n_\sigma - n_{-\sigma}\rangle/\partial h$ with a finite difference $\delta h = 0.005$. If we look at their temperature dependence for various $U \leq 2.0$, χ and χ_0^{RPA} both sharply increase with $T \to 0$, while there is a deviation between them for larger U. The deviation itself indicates that the irreducible four-point vertex $\Gamma = \delta^2\Phi/\delta G\delta G$ cannot be approximated by U. χ_0^{RPA}, which is more divergent for larger U, should be close to reality, since we are dealing with a case where the ferromagnetic phase appears for larger $U(> 6 \sim 7)$, which in turn implies that χ underestimates the effect of external magnetic field for large U.

Now, we come to the result for the magnetoresistance, given in Fig. 17. The figure compares the resistivity against T in the absence and in the presence of

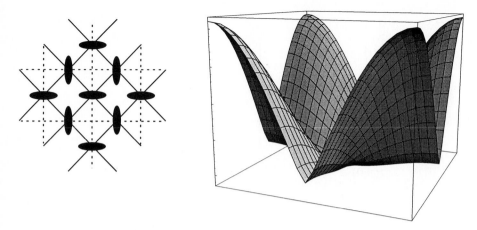

Fig. 18. The in-plane molecular configuration in τ-$D_2A_1A_y$. The solid lines denote the nearest neighbour hopping while dashed lines the second nearest neighbour hopping between face-to-face molecules. The dispersion of the upper band is also depicted.

the magnetic field. We take the system size $N = 64^2$ with 512 Matsubara frequencies, which is checked to be sufficient for the temperature region studied here. We can see that we do have a negative magnetoresistance. The change in the resistance is of the order of 10% for the Zeeman energy $h = 0.05$. If we were not sitting close to the ferromagnetism, much larger fields would be required. The vertex correction does not alter the result significantly, which implies that Boltzmann's transport picture (with no vertex correction) already has the negative magnetoresistance. The negative magnetoresistance should become more prominent at lower temperatures where the ferromagnetic fluctuation increases. This is not too noticeable in the result, which should be an artifact: as mentioned above, the spin polarisation is underestimated in the FLEX at low temperatures for $U = 2$. As touched upon at the beginning, Papavassiliou et $al.$ [65] has found experimentally that organic salts $D_2A_1A_y$, based on D ($=$ P-S, S-DMEDT-TTF or EDO-S,S-DMEDT-TTF) in combination with linear anions A ($=$AuBr$_2$, I$_3$ or IBr$_2$) with the fractional value of y controlling the density of carriers are 2D metals in the τ crystal form. The band structure of a single layer of the τ-phase, calculated with the extended Hückel method, contains a flat-bottomed band [66].

In terms of the tight-binding model, we can regard that the in-plane molecular configuration is such that the next-nearest intermolecular hopping, $t_2 \simeq 0.02\,\mathrm{eV}$, appears in every other plaquetts in a *checker-board* manner on top of the nearest hopping $t_1 \simeq 0.2\,\mathrm{eV}$ (Fig. 18) [67].

The checker-board makes the Brillouin zone folded, where the dispersion of the upper band in which E_F resides is

$$\varepsilon_{\mathbf{k}} = t_2(\cos k_x + \cos k_y) + \sqrt{t_2^2(\cos k_x - \cos k_y)^2 + 4t_1^2(1 + \cos k_x)(1 + \cos k_y)}$$

$$\simeq 2t_1\sqrt{\cos k_x + \cos k_y + \cos k_x \cos k_y + 1},$$

which is just like the square root of Eqn.(22) with $t' = 0.5$ when the splitting due to t_2 is small. This way we can have an anisotropically flat dispersion along k_x, k_y with a large $D(E_F)$ at the bottom of the band, so a ferromagnetic component in the spin fluctuation exists for this dispersion, as we have checked with FLEX. [68]

Murata *et al* [57] have in fact observed a negative magnetoresistance for $B \sim$ several tesla with some hysteretic behaviours in this material [69]. Theoretically, however, we can argue that ferromagnetism is not a necessary condition for the negative magnetoresistance conceived here. In the present scheme, all we have to have about the magnetism is a large enough susceptibility. Thus the negative magnetoresistance observed by Murata *et al* [57] is understandable provided the ferromagnetic *component* is present (which may accompany an antiferromagnetic component) in the spin fluctuation for the present mechanism to be relevant. In fact, the present model for τ-conductor should be strongly antiferromagnetic if t_2 were absent, since the system reduces to a nearly half-filled square lattice. On the other hand ferromagnetic fluctuation is dominant if the lower band is neglected, so antiferromagnetic and ferromagnetic fluctuations should *coexist* for finite values of t_2, which will imply a coexistence of antiferromagnetism and the negative magnetoresistance.

Although we have exemplified our idea here in 2D, we believe that the present mechanism should be general, and can be found in other models with strong ferromagnetic spin fluctuations such as face centred cubic lattice [52], whose band structure has flat and dispersive parts as well.

5   Conclusion

In summary, we have described in this article how in each of the 0D (quantum dots), 1D (quantum wires) and 2D (organic conductors) the effect of electron correlation is exerted and appears in transport properties through the spin degrees of freedom. The 0D physics above exploits the fact that an application of a strong magnetic field makes the electron correlation effect, hence spin effects, stronger in that electrons are spatially localised into Landau's orbits. The 'electron molecule' is a key word there. The 1D physics is a problem of the Tomonaga-Luttinger model, placed in a magnetic field with Zeeman splitting. Degradation of the spin-charge separation is a key word this time. The last item is a novel mechanism for negative magnetoresistance, which is a combined effect of the electron-correlation originated ferromagnetism and the band structure partly flat dispersion). All of these should have, as we have mentioned, relevance to real systems. We can envisage from these examples that the *electron correlation engineering*, so to speak, can be a very promising field for novel electronic properties.

I wish to thank collaborations with Peter Maksym and Hiroshi Imamura on the quantum dot, Takashi Kimura and Kazuhiko Kuroki on the quantum wire and Ryotaro Arita and Kuroki on the organic conductor. I would also like to acknowledge illuminating discussions with Seigo Tarucha and Guy Austing on the quantum dot, Gerhard Fasol and Tetsuo Ogawa on the quantum wire and Keizo Murata, Laurent Ducasse and Hiroshi Kontani on the organic conductor. Some of the works have been supported by a grant-in-aid 'spin controlled semiconductor nanostructures' from the Japanese Ministry of Education.

References

1. See, e.g., *Mesoscopic Phenomena in Solid*, ed. B.L. Altshuler *et al.* (North Holland, 1991); *Transport Phenomena in Mesoscopic Systems*, ed. H. Fukuyama and T. Ando (Springer Verlag, 1991).
2. See, e.g., M.A. Kastner, R.J. Birgeneau, G. Shirane, and Y. Endoh, Rev. Mod. Phys. **70**, 897 (1998). The relation of the spin structure with the shape of the Fermi surface is theoretically discussed by K. Kuroki and H. Aoki, Phys. Rev. B **60**, 3060 (1999); *ibid*, in the press (cond-mat/9904063).
3. See, e.g., S. Das Sarma and A. Pinzcuk (eds): *Perspectives in Quantum Hall Effects* (John Wiley, New York, 1997).
4. H. Imamura, H. Aoki and P.A. Maksym, Phys. Rev. B **57**, R4257 (1998); Physica B **256-258**, 194 (1998).
5. T. Kimura, K. Kuroki and H. Aoki, Phys. Rev. B **53**, 9572 (1996).
6. R. Arita, K. Kuroki and H. Aoki, submitted for publication.
7. D. Weinmann, W. Häusler, and B. Kramer, Phys. Rev. Lett. **74**, 984 (1995); D. Weinmann *et al.*, Europhys. Lett. **26**, 467 (1994).
8. Y. Tanaka and H. Akera, Phys. Rev. B **44**, 3901 (1996).
9. M. Eto, Jpn. J. Appl Phys. **36**, 3924 (1997).
10. S. M. Girvin and T. Jach, Phys. Rev. B **28** (1983) 4506.
11. P.A. Maksym and T. Chakraborty, Phys. Rev. Lett. **65**, 108 (1990).
12. P.A. Maksym, Phys. Rev. B **53**, 10871 (1996); Europhys. Lett. **31**, 405 (1995).
13. H. Aoki, Physica E **1**, 198 (1997).
14. P.A. Maksym, H. Imamura, G. Mallon and H. Aoki, J. Phys. Condens. Matter, to be published.
15. R. J. Galejs, Phys. Rev. B **35**, 6240 (1987).
16. P. A. Maksym, L. D. Hallam and J. Weis, Physica B **212**, 213 (1995).
17. H. Imamura, P.A. Maksym and H. Aoki, Physica B **249-251**, 214 (1998).
18. P.A. Maksym and T. Chakraborty, Phys. Rev. B **45**, 1947 (1992).
19. V. Fock, Z. Phys. **47**, 466 (1928); C.G. Darwin, Proc. Cambridge Philos. Soc. **27**, 86 (1930).
20. To be precise, calculations with a $1/r$ interaction require a larger confinement energy to reproduce experimental results, which is considered to be a consequence of the modification of the interaction potential in real dots (P. A. Maksym and N. A. Bruce, Physica E **1**, 211 (1997)).
21. S. Tarucha, D. G. Austing and T. Honda, Phys. Rev. Lett. **77** 3613 (1996).
22. For $B = 0$ we can make the electron correlation strong if we make the system dilute enough. R. Egger, W. Häusler, C.H. Mak and H. Grabert, Phys. Rev. Lett. **82**, 3320 (1999), have indeed shown recently that a spin blockade can occur even at $B = 0$ in a dot with dilue electrons.
23. H. Imamura, P.A. Maksym and H. Aoki, Phys. Rev. B **52**, 12613 (1996).
24. J. H. Oh *et al.*, Phys. Rev. B **53**, R13264 (1996).
25. J. J. Palacios and P. Hawrylak, Phys. Rev. B **51**, 1769 (1995).
26. J. Hu, E. Dagotto and A. H. MacDonald, Phys. Rev. B **54**, 8616 (1996).

27. D. G. Austing, T. Honda and S. Tarucha, Jpn. J. Appl. Phys. **36**, 11667 (1997).
28. See, e.g., J. Sólyom, Adv. Phys. **28**, 201 (1979); V.J. Emery in *Highly Conducting One-Dimensional Solids*, ed. by J.T. Devreese *et al.* (Plenum, New York, 1979), p.247; H. Fukuyama and H. Takayama in *Electronic Properties of Inorganic Quasi-One Dimensional Compounds*, ed. by P. Monçeau (D. Reidel, 1985), p.41.
29. W. Apel and T.M. Rice, Phys. Rev. B **26**, 7063 (1982); M. Ogata and P.W. Anderson, Phys. Rev. Lett. **70**, 3087 (1993); H. Fukuyama *et al.*, J. Phys. Soc. Jpn **62**, 1109 (1993); A. Kawabata, J. Phys. Soc. Jpn.**63**, 2047 (1994); K.A. Matveev *et al.*, Phys. Rev. Lett. **71**, 3351 (1993); M. Fabrizio *et al.*, Phys. Rev. Lett. **72**, 2235 (1994); N. Nagaosa and A. Furusaki, J. Phys. Soc. Jpn. **63**, 413 (1994).
30. A. Luther and I. Peschel, Phys. Rev. Lett. **32**, 992 (1974) .
31. T. Tarucha, T. Honda and T. Saku, Solid. State. Commun. **94**, 233 (1995).
32. V.M. Edelstein, Solid. State. Commun. **73**, 233 (1990).
33. R. Wiesendanger *et al.*, Phys. Rev. Lett. **65**, 247 (1990).
34. K. Koike *et al.*, Jpn. J. Appl. Phys. **22**, 1332 (1983).
35. G. Fasol and H. Sakaki, Phys. Rev. Lett. **70**, 3643 (1993); G. Fasol and H. Sakaki, Jpn. J. Appl. Phys. **323**, 879 (1994).
36. H. Shiba and M. Ogata, Prog. Theor. Phys. Suppl. **108**, 265 (1992).
37. T. Ogawa and H. Otani, J. Phys. Soc. Jpn. **64**, 3664 (1995); H. Otani and T. Ogawa, Phys. Rev B **54**, 4540 (1996).
38. N. Nagaosa and T. Ogawa, Solid State Commun. **88**, 295 (1993).
39. W. Götze and P. Wölfle, Phys. Rev B **6**, 1226 (1974).
40. T. Kimura *et al.*, Phys. Rev. B **49**, 16852 (1994).
41. For GaAs quantum wires we have typically $n \sim 2 \times 10^{11} \mathrm{cm}^{-2}$, $m^* = 0.067 m_0$, and a single-channel case is realised for the width of the wire $W \sim 400$ A. These values correspond to the cutoff energy $\omega_F \sim 100$K.
42. T. Kimura, K. Kuroki and H. Aoki in *Proc. 23rd Int. Conf. on the Physics of Semiconductors* ed. by M. Scheffler and R. Zimmermann (World Scientific, 1996), p.1225.
43. T. Giamarchi and H.J. Schulz, Phys. Rev. B **37**, 325 (1988).
44. *Numerical Data and Functional Relations in Science and Technology-Crystal and Solid State Physics*, ed. by O. Madelung *et al.*, Landolt-Börnstein, New Series, Group III, Vol. 16, Part a (Springer, Berlin, 1982).
45. H. Kamimura and H. Aoki: *Physics of Interacting Electrons in Disordered Systems* (Oxford Univ. Press, Oxford, 1989).
46. H. Kuwahara and Y. Tokura in *Giant Magnetoresistance and Related Properties of Metal Oxides* ed. by C.N.R. Rao and B. Raveau (World Scientific), in press, and refs therein.
47. J. Kanamori, Prog. Thor. Phys. **30**, 275 (1963).
48. J. Hubbard, Proc. Roy. Soc. London A **276**, 238 (1963).
49. M. C. Gutzwiller, Phys. Rev. Lett. **10**, 159 (1963).
50. R. Hlubina, S. Sorella and F. Guinea, Phys. Rev. Lett. **78**, 1343 (1997).
51. R. Hlubina, Phys. Rev. B **59**, 9600 (1999).
52. R. Arita, S. Onoda, K. Kuroki and H. Aoki, submitted for publication.
53. S. Daul and R. Noack, Z. Phys. B **103**, 293 (1997); Phys. Rev. B **58**, 2635 (1998).
54. R. Arita, Y. Shimoi, K. Kuroki and H. Aoki, Phys. Rev. B, **57** 10609 (1998).
55. R. Arita, K. Kusakabe, K. Kuroki and H. Aoki, Phys. Rev. B, **58** R11833 (1998).
56. R. Arita *et al*, Phys. Rev. B, **57**, R6854 (1998).
57. K. Murata, H. Yoshino, Y. Tsubaki, G. C. Papavasssiliou, Synth. Met. **94**, 69 (1998).

58. R. Arita *et al*, Phys. Rev. B **58**, R11833 (1998).
59. N. E. Bickers, D. J. Scalapino and S. R. White, Phys. Rev. Lett. **62**, 961 (1989); N. E. Bickers and D. J. Scalapino, Ann. Phys. (N. Y.), **193**, 206 (1989).
60. G. Baym and L.P. Kadanoff, Phys. Rev. **124**, 287 (1961); G. Baym, Phys. Rev. **127**, 1391 (1962).
61. H. Kontani, K. Kanki and K. Ueda, Phys. Rev. B **59**, 14723 (1999).
62. G. M. Eliashberg, Sov. Phys. JETP **14**, 886 (1962).
63. K. Yamada and K. Yosida, Prog. Thoer. Phys. **76**, 621 (1986).
64. H. J. Vidberg and J. W. Serene, J. Low. Temp. Phys. **29**, 179 (1977).
65. G. C. Papavasssiliou, D. J. Lagouvardos, A. Terzis, C. P. Raptopoulou, B. Hilti, W. Hofherr, J. A. Zambounis, G. Rihs, J. Pfeiffer, P. Delhaes, K. Murata, N. A. Fortune, and N. Shirakawa, Synth. Met. **70**, 787 (1995).
66. A. Terzis A. Hountas, B. Hilti, C. W. Mayer, J. A. Zambounis, D. J. Lagouvardos, V. C. Kakoussis, G. A. Mousdis and G. C. Papavasssiliou, Synth. Met. **41**, 1715 (1991).
67. L. Ducasse, private communication.
68. If we naively plug in $1/8$ (~ 0.05 eV) of the width of the upper band calculated in ref. [67] for t, the $h = 0.05$ for $U = 2$ in Fig.17 corresponds to $B \simeq 30$ T.
69. A preliminary result (K. Murata, private communications) indicates both an enhanced susceptibility from an M-H curve and antiferromagnetic fluctuations from NMR.

Non-linear Transport in Quantum-Hall Smectics

A.H. MacDonald and Matthew P.A. Fisher

[1] Department of Physics, Indiana University, Bloomington, IN 47405-4202
[2] Institute for Theoretical Physics, University of California at Santa Barbara, Santa Barbara CA 93106-4030

Abstract. Recent transport experiments have established that two-dimensional electron systems with high-index partial Landau level filling, $\nu^* = \nu - [\nu]$, have ground states with broken orientational symmetry. In a mean-field theory, the broken symmetry state consists of electron stripes with local filling factor $[\nu] + 1$, separated by hole stripes with filling factor $[\nu]$. We have recently developed a theory of these states in which the electron stripes are treated as one-dimensional electron systems coupled by interactions and described by using a Luttinger liquid model. Among other things, this theory predicts non-linearities of opposite sign in easy and hard direction resistivities. In this article we briefly review our theory, focusing on its predictions for the dependence of non-linear transport exponents on the separation d between the two-dimensional electron system and a co-planar screening layer.

1 Introduction

Recent transport experiments [1–3] have established that the resistivity of a two-dimensional electron system with weak disorder and valence orbital Landau level filling ν^* close to $1/2$ is anisotropic when the valence orbital Landau level index $N \geq 2$. Apparently the ground state spontaneously breaks orientational symmetry, a property believed to be associated with the uni-directional charge-density-wave ground states predicted [4,5] in this regime by Hartree-Fock mean-field theory [7]. The charge-density-wave (CDW) state consists of electron stripes of width $a\nu^*$ with local Landau level filling factor $\nu = [\nu] + 1$, separated by hole stripes of width $a(1 - \nu^*)$ with local filling factor $\nu = [\nu]$. Here a, the CDW period, is comparable to the Landau level's cyclotron orbit diameter. Because of symmetry properties shared with the smectic state of classical liquid crystals, emphasized by Fradkin and Kivelson [6], these states have been referred to as quantum Hall smectics, a practice we follow here. The most important transport property of quantum Hall smectics is that dissipation occurs over a wide range of filling factors surrounding $\nu^* = 1/2$ and is not activated at low temperatures. This behavior is *not* consistent with the properties of a CDW state, which would be pinned by the random disorder potential and have a large gap for mobile quasiparticle excitations, and suggests that, although Hartree-Fock theory hints at the energetic motivation for a ground state with broken orientational symmetry, the description which it provides of the ground state is flawed.

Several recent theoretical papers [8–13] have addressed the properties of quantum Hall smectics and the energetic competition between CDW states, compressible composite-fermion states, and paired incompressible quantum Hall states. In one recent paper [14] we have described a theory of quantum Hall smectics which starts from the Hartree-Fock theory ground state, recognizes the electron stripes as coupled one-dimensional electron systems, and treats residual interaction and disorder terms neglected by Hartree-Fock theory using the convenient bosonization techniques of one-dimensional electron physics. The most important conclusions of this work are the following: i) the quantum Hall smectic state is *never* the ground state but instead is always unstable, for ν^* close to 1/2 likely to an anisotropic electron Wigner crystal state; ii) for $0.4 \lesssim \nu^* \lesssim 0.6$ the interaction terms responsible for the Wigner crystal can be neglected at temperatures available in a dilution fridge; and iii) weak disorder which scatters electrons from stripe to stripe, enabling hard-direction transport, leads to non-linear transport. In this article we emphasize and expand on an experimentally important prediction of this work, namely that the strength of the transport non-linearity is sensitive to the nature of the electron-electron interaction. In particular we predict that the transport non-linearity can be enhanced by placing a screening plane close to the two-dimensional electron system.

In Section 2 we explain our theory of transport in quantum Hall smectics and discuss how the coefficients which govern the power-law behavior of the differential resistivity for weak disorder are related to correlations in the coupled one-dimensional electron stripes. In Section 3 we briefly review our theory of these correlations and explain why long-range electron-electron interactions weaken transport non-linearities. In Section 4 we present numerical results for the non-linear transport coefficients for a model in which interactions in the two-dimensional electron layer are screened by a metallic layer co-planar with the electron system but separated from it by a distance d. We conclude in Section 5 with a brief summary.

2   Anisotropic Transport Properties

Our transport theory [14] is built on a semiclassical Boltzmann-like approach; microscopic physics enters only in the calculation of scattering rates. We choose a coordinate system where the $\hat{x}$ (horizontal) direction runs along the stripes which are separated in the $\hat{y}$ (vertical) direction. We assume that the charge density wave itself is pinned and immobilized by both the edges of the sample and weak impurities which couple to the electrons within the stripes. In this case, collective sliding motion of the charge-density will be absent, and electrical transport will be dominated by single-electron scattering across and between electron stripes. An important property of electronic states in the quantum Hall regime, is the spatial separation of states which carry current in opposite directions. In the case of the electron stripes in quantum Hall smectics, the Fermi edge states which carry oppositely directed currents along the stripes (left-going and right-going) are located on opposite sides of the stripe. Translational invariance along the $\hat{x}$ direction, allows us to use a Landau gauge where this component of wavevector $\hbar k_x$ is a good quantum number in the absence of interactions and disorder. The single-particle states at the stripe edges have velocity magnitude v_F, the Fermi velocity. In the Landau gauge, $\hat{x}$ direction momenta are related to $\hat{y}$ direction positions by $k_x = y/\ell^2$ where $\ell \equiv (\hbar c/eB)^{1/2}$ is the magnetic length, so that states on opposite sides of the same stripe differ in momenta by a $\nu^* a/\ell^2$ and the adjacent sides of neighboring stripes differ in momenta by $(1 - \nu^*)a/\ell^2$.

We now summarize the basic assumptions on which our semiclassical transport theory is based and quote the expressions implied by these assumptions. For further details see Ref. [14]. We assume that in the steady state, each edge of each stripe is characterized by a local chemical potential. Translational invariance in the $\hat{y}$ direction implies that the chemical potential drops across each stripe and between any two adjacent stripes are the same, mf μ is the chemical potential drop across an electron stripe, it follows that the potential drop between stripes is $eE_y a - \mu$, where E_y is the hard-direction electric field, the field which supports a steady state transport across the stripes. The electric field in the $\hat{x}$ direction produces a semiclassical drift in momentum space which drives the system from equilibrium. We assume that disorder scattering across and between stripes, then attempts to reestablish equilibrium and that the drift and scattering processes are in balance in the steady state. The scattering currents in the hard-direction are characterized by relaxation times τ_e and τ_h respectively. The current along a stripe is analogous to the quantized current in a long narrow Hall bar and is proportional to the chemical potential difference across that stripe. Combining these ingredients leads [14] to the following expressions for the resistivities:

$$\rho_{\text{easy}} = \frac{h}{e^2} \frac{1}{\tau_e([\nu] + 1)^2 + \tau_h[\nu]^2} \frac{a}{v_F}$$

$$\rho_{\text{hard}} = \frac{h}{e^2} \frac{1}{\tau_e([\nu] + 1)^2 + \tau_h[\nu]^2} \frac{v_F \tau_e \tau_h}{a}$$

$$\rho_{\text{hall}} = \frac{h}{e^2} \frac{1}{\tau_e([\nu] + 1)^2 + \tau_h[\nu]^2} ([\nu] + 1)\tau_e + [\nu]\tau_h, \tag{1}$$

where $\rho_{\text{easy}} = \rho_{xx}$, $\rho_{\text{hard}} = \rho_{yy}$, and $\rho_{\text{hall}} = \rho_{xy}$.

For $\nu^* = 1/2$, this theory makes a parameter free prediction for the product $\rho_{easy}\rho_{hard}$ which has been confirmed experimentally [15]. In fact, as emphasized [13] by van Oppen et al., this feature of our results has a greater validity than would be suggested by our assumption of largely intact electron stripes. Our main interest here however, is in expanding on our predictions [14] for non-linearities in the easy and hard direction differential resistivities. These predictions were made on the basis of a simple lowest order renormalization group scheme for handling the infrared divergence which appear when disorder terms which scatter electrons either across or between stripes are treated perturbatively. This analysis leads to

$$\frac{1}{\tau_e} \equiv \Gamma_e \sim \Gamma_e^{(0)}(V_y/E_c)^{2\Delta_e - 2}$$

$$\frac{1}{\tau_h} \equiv \Gamma_h \sim \Gamma_h^{(0)}(V_y/E_c)^{2\Delta_h - 2} \tag{2}$$

where $\Gamma_e^{(0)}$ and $\Gamma_h^{(0)}$ are Golden-rule scattering rates at the characteristic microscopic energy scale E_c, Here Δ_e is [14] the scaling dimension of the operator which scatters an electron across a stripe, which we discuss in the next section, and Δ_h is the scaling dimension of the operator which scatters an electron between neighboring stripes. The values of Δ_e and Δ_h depend on correlations induced by electron-electron interaction between stripes, and are sensitive in particular to the range of the microscopic electron-electron interaction. At $\nu^* = 1/2$, the case on which we will concentrate, $\Delta_e = \Delta_h$.

Given these expressions, it follows [14] that the non-linear differential resistivity in the hard direction

$$\frac{\partial V_y}{\partial I_y} \sim I_y^\alpha, \tag{3}$$

with an exponent $\alpha = 2(1 - \Delta_e)/(2\Delta_e - 1)$. Similar considerations apply for the easy direction current:

$$\frac{\partial V_x}{\partial I_x} \sim I_x^\beta \tag{4}$$

with an exponent $\beta = 2(\Delta_e - 1)$. In the next section we show that both α and β increase when the distance to the screening plane is comparable to or smaller than the CDW period.

3   Quantum Smectic Model

The CDW state of Hartree-Fock theory [4,5] is a single-Slater-determinant. In the valence Landau level, groups of Landau-gauge single-particle states with adjacent k_x (adjacent $\hat{y}$) are occupied to form stripes and separated by groups which are unoccupied. Small fluctuations in the positions and shapes of the stripes can be described in terms of particle-hole excitations near the stripe edges. The residual electron-electron interaction terms which scatter into these low energy states are ignored in Hartree-Fock theory and fall into two classes: "forward" scattering interactions which conserve the number of electrons on each edge of every stripe, and "backward" scattering processes which do not. The latter processes involve large momentum transfer and are unimportant [14] at accessible temperatures for ν^* near 1/2. The quantum smectic model [14], briefly described in this section includes forward scattering only. The interactions are bilinear in the 1D electron density contributions from a particular edge of a particular stripe: $\rho_{n\alpha}(x)$, with $\alpha = \pm$. Since the density of a single filled Landau level is $(2\pi\ell^2)^{-1}$, the displacement of an edge is related to its associated charge density contribution by $u_{n\alpha}(x) = \alpha 2\pi\ell^2 \rho_{n\alpha}(x)$. The quadratic Hamiltonian which describes the *classical* energetics for small fluctuations has the following general form:

$$H_0 = \frac{1}{2\ell^2} \int_{x,x'} \sum_{n,n'} u_{n\alpha}(x) D_{\alpha\beta}(x - x'; n - n') u_{n'\beta}(x')$$

$$= \frac{1}{2\ell^2} \int_q u_\alpha(-\mathbf{q}) D_{\alpha\beta}(\mathbf{q}) u_\beta(\mathbf{q}), \tag{5}$$

where $\int_q \equiv \int d^2\mathbf{q}/(2\pi)^2$. Here the q_y integral is over the interval $(-\pi/a, \pi/a)$ and a high momentum cutoff $\Lambda \sim 1/\ell$ is implicit for q_x.

Symmetry considerations constrain the form of the elastic kernel. In position space, the kernel must be real and symmetric so that $D_{\alpha\beta}(\mathbf{q}) = D_{\alpha\beta}^*(-\mathbf{q}) = D_{\beta\alpha}^*(\mathbf{q})$. This implies $D_{-+}(\mathbf{q}) = D_{+-}^*(\mathbf{q})$ and $\mathrm{Im}D_{\alpha\alpha}(\mathbf{q}) = 0$. Parity invariance (under $x, n, + \leftrightarrow -x, -n, -$), implies moreover $D_{++}(\mathbf{q}) = D_{--}(\mathbf{q})$. Thus, the elastic kernel is fully specified by one real function, $D_{++}(\mathbf{q})$, and one complex function, $D_{+-}(\mathbf{q})$. It will be important for our present interest that the Hamiltonian must be invariant under: $u_{n\alpha}(x) \to u_{n\alpha}(x) + const$. For short-range interactions this implies that at long wavelengths

$$D(q_x = 0, q_y) = K_y q_y^2 + ..., \tag{6}$$

characteristic of classical smectic elasticity. As we will discuss, this conclusion must be modified in the case of long-range interactions.

A *quantum* theory of the Quantum-Hall smectic [14] is obtained by imposing Kac-Moody commutation relations on the chiral densities:

$$[\rho_{n\alpha}(x), \rho_{n'\beta}(x')] = \frac{i}{2\pi}\alpha\delta_{\alpha,\beta}\delta_{n,n'}\partial_x\delta(x - x'). \tag{7}$$

Together with Eq.(5), this relationship fully specifies the quantum dynamics. Electron operators in the chiral edge modes are related to the 1D densities via the usual bosonic phase fields: $\psi_{n\alpha} \sim e^{i\phi_{n\alpha}}$ with $\rho_{n\alpha} = \alpha\partial_x\phi_{n\alpha}/2\pi$.

Quantum properties of the smectic can be computed from the imaginary-time action,

$$S_0 = \int_{x,\tau} \frac{1}{4\pi} \sum_{n,\alpha} i\alpha\partial_\tau\phi_{n,\alpha}\,\partial_x\phi_{n,\alpha} + \int_\tau H_0$$
$$= \frac{1}{2}\int_{\mathbf{q},\omega} \phi_\alpha(-\mathbf{q}, -\omega)M_{\alpha,\beta}(\mathbf{q},\omega)\phi_\beta(\mathbf{q},\omega), \tag{8}$$

where in an obvious matrix notation,

$$\mathbf{M}(\mathbf{q},\omega) = (i\omega q_x/2\pi)\sigma^z + (q_x\ell)^2\mathbf{D}(\mathbf{q}). \tag{9}$$

Correlation functions follow from Wick's theorem and the momentum space correlator $\langle\phi_\alpha\phi_\beta\rangle = \mathbf{M}^{-1}$ with

$$\mathbf{M}^{-1}(\mathbf{q},\omega) = \sigma_z\mathbf{M}(\mathbf{q},-\omega)\sigma_z/\det\mathbf{M}(\mathbf{q},\omega). \tag{10}$$

The effect of weak disorder on transport in quantum Hall smectics depends sensitively on the elastic constants *at* $q_x = 0$. In this limit the relevant excited states are simply Slater determinants with straight stripe edges displaced from those of the Hartree-Fock theory ground state. By evaluating the expectation value of the microscopic Hamiltonian in a state with arbitrary stripe edge locations we find that

$$D_{\alpha\beta}(q_x = 0, q_y) = \delta_{\alpha\beta}D_0 + \alpha\beta\frac{a}{4\pi^2\ell^2}\sum_n e^{iq_y an}\Gamma(y_{n\alpha}^0 - y_{0\beta}^0), \tag{11}$$

where the constant D_0 is such that the condition $\sum_{\alpha\beta} D_{\alpha\beta}(\mathbf{q}=0)=0$, and the positions $y_{n\pm}^0 = a(n \pm \nu^*/2)$ are the ground state stripe edge locations. Here, $\Gamma(y)$ is the interaction potential between two electrons located in guiding center states a distance y apart:

$$\Gamma(y) = U(0, y/\ell^2) - U(y/\ell^2, 0),\tag{12}$$

$$U(q,k) = \int \frac{dp}{2\pi} e^{-(q^2+p^2)\ell^2/2} V_{\text{eff}}^N(q,p) e^{-ipk\ell^2}.\tag{13}$$

The two terms in Eq. (12) are direct and exchange contributions. In Eq. (13), $V_{\text{eff}}^N(q,p)$ is the Fourier transform of the effective 2D electron interaction which incorporates form-factors [11] dependent on the Landau level index. N and the ground subband wavefunction of the host semiconductor heterojunction or quantum well. The smectic states have relatively long periods proportional to the cyclotron orbit radii. Explicit calculations [4,10,11] show that $a \gtrsim 6\ell$ for $N = 2$. It follows that the exchange contribution to $\Gamma(y)$ is small and that $\Gamma(y)$ decreases with stripe separation in the relevant range. In this paper we address the influence of a metallic screening plane which cuts off this interaction at large distances. For $y \lesssim d$, $\Gamma(y) \sim 2e^2 \ln(d/y)$, decreasing extremely slowly with y. For separation y larger than the distance d to the screening plane, $\Gamma(y) \sim y^{-2}$, making the sum over n in Eq. 11 convergent.

4   Screening Dependence of Scaling Dimensions

The scaling dimension, Δ_e, of the operator $e^{i(\phi_{n,+}-\phi_{n,-})}$ which scatters an electron across the n-th stripe is readily evaluated from Eq.(9). We find that

$$\Delta_e = \int_{-\pi}^{\pi} \frac{d(qa)}{2\pi} W(q_x = 0, q).\tag{14}$$

Here, W is the weight function,

$$W(\mathbf{q}) = \frac{[D_{++}(\mathbf{q}) + \text{Re}D_{+-}(\mathbf{q})]}{[D_{++}^2(\mathbf{q}) - |D_{+-}(\mathbf{q})|^2]^{1/2}}.\tag{15}$$

For 1D non-interacting electrons $\Delta_e = 1$, so that disorder is relevant and eventually leads to localization. As discussed below, $\Delta_e < 1$ for quantum Hall smectics. Disorder is even more relevant than in the non-interacting electron case. Nevertheless, since the samples in which the quantum Hall smectic is observed are of extremely high quality, there should be a wide range of temperature over which its effects can be treated perturbatively. If $\Delta_e = 1$ both hard direction (α) and easy direction (β) non-linear transport exponents vanish. We see from Eq.(15) that $\Delta_e = 1$ if the average value of $W(\mathbf{q_y})$ is one. To understand the dependence of Δ_e on screening, we have to understand the dependence of $W(q_y)$ on both wavevector and d.

Note that W is smaller than one, increasing the relevance of disorder, when D_{++} and $\mathrm{Re}D_{+-}$ are opposite in sign and similar in magnitude. For each q_y in Eq. 14 the weighting factors are like those which enter in the calculation of the scaling dimension of the operator which describes backscattering from disorder in an isolated one-dimensional electron system. In continuum 1D models, D_{++} has a contribution, proportional to the Fermi velocity, from the band energy and a contribution proportional to the interaction between electrons traveling in the same direction, while $-D_{+-}$ has only an interaction contribution. For a continuum model, the effective interactions between electrons traveling in different directions is the same as that between electrons traveling in the same directions. When the interaction term is much larger than the band term D_{++} and D_{+-} are opposite in sign and nearly equal in magnitude and W is very small. This is what happens, for example, for a 1D electron system in which long-range makes the Coulomb interaction very strong at long wavelengths. When W is small, the 1D electron system is very close [16] to an electron Wigner crystal, and disorder is very strongly relevant. On the other hand when D_{+-} is much smaller than D_{++} we have a situation analogous to that in a very weakly interacting Fermion system, in which disorder is relevant but the resistivity is linear when disorder can be treated perturbatively.

With this in mind we turn to a discussion of the quantum smectic, limiting our attention to the case $\nu^* = 1/2$. Useful insight comes from examining the value of W at the end points of the integration interval, $q_y a = 0$ and $q_y a = \pi$ where both D_{++} and D_{+-} are real. For $q_y = 0$, invariance under a uniform translation of the smectic implies that $D_{++} + D_{+-} = 0$, so that $W(q_y = 0) = 0$. When all the electron stripes move together, the energetics is precisely like that of a single 1D system. In the quantum Hall regime, there is no band energy, only interaction contributions from electrons traveling in the same direction, which appear in D_{++}, and interaction contributions from electrons traveling in opposite directions, which appear in D_{+-}. The absence of a band contribution means that, for $q_y = 0$, W vanishes independent of the interaction's strength or range. When $q_y a = \pi$, on the other hand, one has

$$D_{+-}(q_y a = \pi) = \sum_n \frac{(-1)^n a}{4\pi^2 \ell^2}[\Gamma(an + a(1 - \nu^*)) - \Gamma(an + a\nu^*)]. \qquad (16)$$

We see that $D_{+-}(q_y a = \pi)$ vanishes because the interaction between the top of one stripe and the bottom of the same stripe is identical to its interaction with the bottom of the next stripe up. The interactions between oppositely directed electrons effectively cancel out, and we obtain $W = 1$, just as we would for a non-interacting 1D system. Thus Δ_e is determined by the average over q_y of a weighting factor which interpolates between that of a 1D electron Wigner crystal at $q_y a = 0$ and that of a non-interacting 1D electron system at $q_y a = \pi$.

The average value of W is determined by the rate at which W goes from its $q_y = 0$ limit to its $q_y = \pi$ limit. To obtain insight into what controls this, we consider first the limit of short-range interactions. Since $\Gamma(y = 0)$ vanishes due to the cancellations of its direct and exchange contributions, the short-range limit is obtained by taking only $\Gamma(a/2) \equiv \Gamma^* \neq 0$. In this case, it follows from Eq.(11) that $(4\pi^2\ell^2)D_{+-}(q)/a = -\Gamma^*(1+\exp(-iqa))$ and that $(4\pi^2\ell^2 D_{++}(q)/a = 2\Gamma^*$, and therefore that $W(q) = |sin(qa/2)|$. The numerator of Eq.(15) vanishes like q^2 for $q \to 0$, and the denominator, which is proportional to the collective mode velocity, vanishes like $|q|$. Note that the expression for $W(q)$ is independent of Γ^*. The average of $W(q)$ may be evaluated analytically in this case, and we obtain for short-range interactions $A_e = 2/\pi \approx 0.6366$.

For the realistic case, analytic calculations are no longer possible, but the behavior of W can be simplified, at least at small q if the exchange contribution to $\Gamma(y)$ is neglected in constructing $D_{\alpha,\beta}(q)$. In this case $\Gamma(y)$ is the simply the 1D transform to coordinate space of the reciprocal space interaction $U(p) \equiv \exp(-p^2\ell^2/2)V_{eff}^N(0,p)$, i.e. it is the Coulomb interaction between lines of charged smeared by N-dependent cyclotron orbit form factors. The components of $D_{\alpha,\beta}(q)$, are then Fourier transforms back to reciprocal space, but with additional 'umklapp' terms because this transform is discrete. We find that

$$D_{+-}(q) = \frac{-1}{4\pi^2\ell^2} \exp(-iqa/2) \sum_{j=-\infty}^{\infty} (-)^j U(q+2\pi j/a)$$

$$D_{++}(q) = D_0 + \frac{1}{4\pi^2\ell^2} \sum_{j=-\infty}^{\infty} U(q+2\pi j/a) \quad (17)$$

and

$$D_0 = \frac{1}{4\pi^2\ell^2} \sum_{j=-\infty}^{\infty} U(2\pi j/a)[(-)^j - 1]. \quad (18)$$

For the case of a Coulomb interaction, $U(q) = [2\pi e^2(1 - \exp(-2qd))]/q$ where d is the distance to a screening plane described by an image charge model. For large or infinite d, the $j = 0$ terms dominate the sums in Eqs. (17). The above expression for $U(q)$ applies when $q\ell \lesssim 1$, and therefore is always valid for the $j = 0$ terms in Eqs.(17). Note that both $D_{++}(q)$ and $D_{\pm}(q)$ are proportional to d for large d and diverge for $d \to \infty$. On the other hand D_0 remains finite for $d \to \infty$ because the $j = 0$ term is excluded from this sum. We emphasize that the numerical calculations whose results are shown below include the exchange contributions neglected in deriving Eqs.(17), and important at any value of a when d is not large.

We now examine the large d, small q behavior of the numerator and denominator of Eq.(15). For the numerator

$$D_{++}(q) + \mathrm{Re}D_{+-}(q) = \frac{1}{4\pi^2\ell^2}U(q)(1 - cos(qa/2)) \sim \frac{e^2}{2\pi\ell^2}q^2 da^2/2, \quad (19)$$

and for the denominator

$$D_{++}(q)^2 - |D_{+-}(q)|^2 \propto U(q) \propto [d]\,[q^2a^2]. \quad (20)$$

The first factor is square brackets in Eq.(20) comes from $D_{++}(q)+|D_{+-}(q)|$ for which the $j=0$ term dominates. The second factor is proportional to $D_{++}(q)-|D_{+-}(q)|$ for which only odd j terms survive, implying no dependence on d and analytic dependence on q. These formulas apply for $qd < 1$; for $d \to \infty$, the small q behavior is obtained by replacing $U(q = 0) = 4\pi e^2 d$ by $U(q \to 0) = 2\pi e^2/q$, i.e., by replacing d by $1/2q$. We plot the square of the denominator, proportional to the square of the collective excitation velocity for several values of the screening length d in Fig. 1. The velocity increases as d increases as expected. For $d = 100\ell$, the quadratic small q behavior predicted in Eq.(20), applies only for $qa \lesssim 0.02$, and is not apparent in the plot. Instead we see the long range interaction behavior, in which the velocity is proportional to $q^{1/2}$. As is apparent from Eqs.(17), screening is irrelevant except very close to $q = 0$

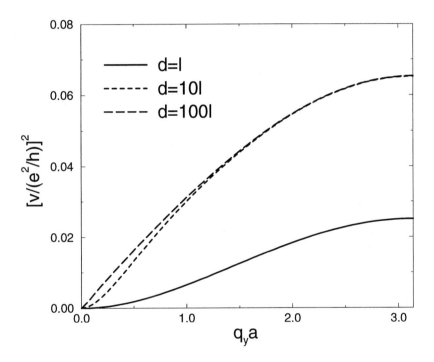

Fig. 1. Square of the collective excitation velocity (in units of $e^2/(2\pi\epsilon\hbar)$) in the $\hat{x}$ direction as a function of q_y for $d = \ell$, $d = 10\ell$ and $d = 100\ell$. These calculations are for $N = 2$ and CDW period $a = 5.8\ell$. For the dielectric function of GaAs, this velocity unit has the value $4.8 \times 10^4\,\mathrm{m/s}$.

when d is large. Fig. 1 shows that once $d \gtrsim a$, the large d no-screening limit is approached. In Fig. 2 we show the weight functions for $d = \ell$, $d = 10\ell$ and $d = 100\ell$. At each value of d, the denominator of the weight function at small

q is proportional to $d^{1/2}|q|$ and the numerator proportional to dq^2. The weight function is therefore proportional to $|q|$ with a coefficient which varies as $d^{1/2}$. A larger value of d (less screening), leads to a weight function which increases more rapidly with q and a scaling dimension for the scattering vertex which is closer to one. In the limit of unscreened interactions, which applies down to small q for $d = 100\ell$, $W(q) \propto |q|^{1/2}$

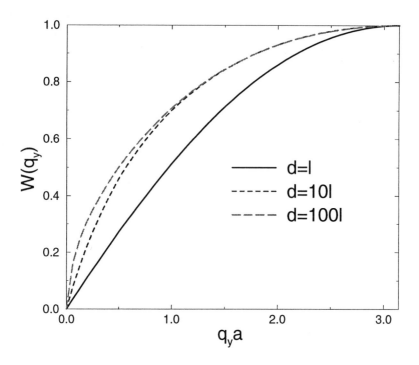

Fig. 2. Weight function *vs.* $q_y a$ for $d = \ell$, $d = 10\ell$ and $d = 100\ell$. These calculations are for $N = 2$ and CDW period $a = 5.8\ell$.

In Fig. 3 we show the dependence of the scaling dimension Δ_e and the non-linear transport exponents α and β on the distance d to the screening plan. For $d \to 0$ the numerical result is very close to that from the analytic short-range interaction model described above which leads to $\Delta_e = 2/\pi$. For $d \to 0$, the scaling dimension approaches $\Delta_e = 0.772$, a value we have been able to obtain only numerically. These relatively modest changes in the scaling dimension translate into relatively large changes in the transport exponents, particularly in α which characterizes the hard-axis non-linearity. We predict that these non-linearities will be much stronger if a screening placed in close proximity to the electron layer.

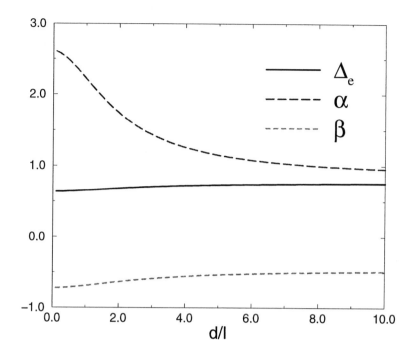

Fig. 3. Scaling dimension Δ_e and non-linear transport exponents α and β as a function of d/ℓ. These calculations are for $N = 2$ and CDW period $a = 5.8\ell$.

5   Summary

Recent experiments [1–3] have established a consistent set of transport properties for high-mobility two-dimensional electron systems with high orbital index ($N \geq 2$) partially filled Landau levels. These properties differ qualitatively from those which occur in the low orbital index ($N \leq 1$) fractional quantum Hall effect regime. At large N, the dissipative resistivities are large, strongly anisotropic, and non-linear for $0.4 \lesssim \nu - [\nu] \lesssim 0.6$ within each Landau level. This anisotropic transport regime is bracketed by regions of reentrant integer quantum Hall plateaus. We have recently [14] developed a theory which is able to account for most features of these experiments. An important prediction of the theory is that the easy and hard direction resistivities should have non-integral power-law temperature dependences. In this article we have briefly summarized the theory and elaborated on its predictions for the dependence of these exponents on the distance d between the two-dimensional electron layer and a remote screening plane, predictions which are summarized in Fig. 3. We find that Δ_e approaches two different values, both smaller than one, for $d \to 0$ and $d \to \infty$, and interpolates smoothly between these limits at finite values of d. Verification of our prediction that non-linear transport can be enhanced by introducing a screening plane and reducing d, would help substantiate our theory of quantum Hall smectics.

Acknowledgements

We would like to acknowledge insightful conversations with Herb Fertig, Michel Fogler, Tomas Jungwirth, Jim Eisenstein, Eduardo Fradkin, Steve Girvin, and Philip Phillips. AHM acknowledges the hospitality of the ITP at UC Santa Barbara, where this work was initiated. M.P.A.F. is grateful to the NSF for generous support under grants DMR-97-04005, DMR95-28578 and PHY94-07194. A.H.M. is grateful for support under grant DMR-97-14055.

References

1. M.P. Lilly, K.B. Cooper, J.P. Eisenstein, L.N. Pfeiffer, and K.W. West, Phys. Rev. Lett. **82**, 394 (1999); *ibid* **83**, 824 (1999).
2. R.R. Du, D.C. Tsui, H.L. Störmer, L.N. Pfeiffer, K.W. Baldwin, and K.W. West, Solid State Commun. **109**, 389 (1999); W. Pan, R.R. Du, H.L. Störmer, D.C. Tsui, L.N. Pfeiffer, K.W. Baldwin, and K.W. West, Phys. Rev. Lett. **83**, 820 (1999).
3. M. Shayegan and H.C. Manoharan, preprint [cond-mat/9903405] (1999).
4. A.A. Koulakov, M.M. Fogler, and B.I. Schklovskii, Phys. Rev. Lett. **76**, 499 (1996); Phys. Rev. B **54**, 1853 (1996); M.M. Fogler and A.A. Koulakov, Phys. Rev. B **55**, 9326 (1997).
5. R. Moessner, and J.T. Chalker, Phys. Rev. B **54**, 5006 (1996).
6. E. Fradkin and S.A. Kivelson, Phys. Rev. B **59**, 8065 (1999).
7. This expectation is based on Hartree-Fock theory which predicts charge-density-wave ground states throughout the quantum Hall regime, usually incorrectly. See for example A.H. MacDonald and D.B. Murray, Phys. Rev. B **32**, 2291 (1985). The Hartree-Fock predictions of broken symmetry states are more plausible in higher Landau levels, however, because the associated variational energies compete favorably with variational estimates of fluid state energies.
8. E.H. Rezayi, F.D.M. Haldane, and Kun Yang, Phys. Rev. Lett. **89**, 1219 (1999).
9. H.A. Fertig, Phys. Rev. Lett. **82**, 3693 (1999).
10. Tudor Stanescu, Ivar Martin and Philip Phillips, preprint [cond-mat/9905116] (1999) - to appear in Phys. Rev. Lett..
11. See for example T. Jungwirth, A.H. MacDonald, and S.M. Girvin, Phys. Rev. B **60**, 15574 (1999).
12. Ziqiang Wang, preprint [cond-mat/9911265] (1999).
13. Felix von Oppen, Bertrand I. Halperin and Ady Stern, preprint [cond-mat/9910132] (1999).
14. A.H. MacDonald and M.P.A. Fisher, preprint [cond-mat/9907278] to appear in Phys. Rev. B.
15. M.P. Lilly and J.P. Eisenstein, private communication (1999).
16. H.J. Schulz, Phys. Rev. Lett. **71**, 1864 (1993).

Thermodynamics of Quantum Hall Ferromagnets

Marcus Kasner

Institut für Theoretische Physik, Otto-von-Guericke-Universität Magdeburg,
PF 4120, D-39016 Magdeburg, Germany

Abstract. A two-dimensional electron system in a strong magnetic field exhibits at certain filling factors ν a non-zero spin magnetization in the ground state. When this even persists for vanishing Zeeman coupling the system is called a quantum Hall ferromagnet. This happens at $\nu = 1$ and $\nu = 1/3$, but also around these filling factors. For $\nu = 1$ we show that the interaction between the electrons is a necessary ingredient of any theory in order to understand the temperature dependent spin magnetization as an example of a thermodynamic quantity. We develop a theory that takes into account spin-wave corrections to the electronic self-energy by going beyond effectively one-particle theories. This leads to an improved magnetization when compared with experimental data. Furthermore, we compare our findings with other theoretical approaches.

1 Introduction

When electrons in a two-dimensional electron system (2DES) like in *GaAs*-heterostructures are subject to a constant, perpendicular magnetic field $\boldsymbol{B}$ the one-particle electronic states form orbital Landau levels $n = 0, 1, \ldots$ each splitting up into spin-up ($\uparrow$) and spin-down energy levels ($\downarrow$). This is due to the coupling of the electron's spin to the magnetic field via a non-vanishing g-factor in the Zeeman term of the Hamiltonian. Each of these levels (n, σ) has a degeneracy that is given by the number of elementary flux quanta $N_\Phi = BA/\Phi_0$ ($\Phi_0 = h/e$) through the sample area A so that the filling factor $\nu = N/N_\Phi$ (N - number of electrons) describes the degree of filling of the levels' one-particle states at zero temperature. In *GaAs*-heterostructures we encounter the situation that the ratio of the Zeeman energy to the energy between adjacent Landau levels $\Delta_z/(\hbar\omega_c)$ ($\Delta_z = |g\mu_B B|, \omega_c = |eB|/m^*$, where μ_B is Bohr's magneton and m^* is the effective mass) is about $1/60$ instead of one as in the vacuum. Therefore the energy levels are clearly separated by their Landau levels and it is justified to neglect the physics of higher Landau levels with $n \geq 1$ for sufficiently high magnetic fields. Such an assumption is even good when for the experimentally accessible strong magnetic fields the interaction between the electrons is taken into account.

The characteristic transport properties at high magnetic fields at temperatures at about and below 1 K can be observed in the integer and fractional quantum Hall effects and define the quantum Hall regime. Current progress in the development of experimental methods and in sample preparation allowed to extend the study of 2DES beyond transport measurements. Before, interest was mostly focussed on the study of the low temperature region and was less aimed to the investigation of thermodynamic quantities covering the whole temperature range. Now, experimental data for the temperature dependent spin magnetization from nuclear magnetic resonance (NMR) [1] and magneto-optical absorption experiments [2], for the total magnetization including the orbital contribution from torque magnetometer [3] and SQUID experiments [4] and also for the specific heat [5] are available at certain filling factors in the quantum Hall regime. This opens the unique opportunity to compare theoretical results directly with experimental data in systems which are, on one hand, exceptionally clean due to its high mobility and where, on the other hand, electronic correlation plays an essential role. Since the Landau levels are dispersionless comparison with experiment is not hampered as in the case of other strongly correlated systems where the peculiarities of the band structure have to be incorporated in any satisfying theory.

It is instructive to start with a general consideration of the ground state spin magnetization of interacting electrons in two dimensions in dependence on the filling factor ν. Because the Zeeman coupling Δ_z can be viewed as a symmetry breaking field the existence of a non-vanishing magnetization of the electronic system for zero Zeeman coupling represents spontaneous magnetization. Due to the continuous $SU(2)$-symmetry of the interaction term the Mermin-Wagner theorem applies and only at zero temperature the spontaneous magnetization can survive, while for any finite temperature the spin magnetization has to vanish in two dimensions. This suggests the notion of a *quantum Hall ferromagnet*, when the ground state is ferromagnetic. In this respect quantum Hall ferromagnets are similar to three-dimensional metallic ferromagnets without external magnetic field. Examples for such ground states in a 2DES can be found for filling factors exactly at $\nu = 1$, about $\nu = 1$ [1] and at 1/3 [6], while for some fractional ν like 2/5 the ground state forms for $\Delta_z = 0$ a spin-singlett and is therefore paramagnetic. In real samples the Zeeman energy is small, but does not vanish and cuts off for any finite Δ_z the decrease of the magnetization which leads to a non-trivial magnetization curve over the whole temperature region.

It should be noted that the electronic system with its two spin orientations along the z-axis can be viewed as a realization of a multi-component electron system, which can be found,e. g. , in a system of parallel layers each forming a 2DES with spin-polarized electrons. The layer index can be identified as a pseudo-spin, so that in the somewhat artificial situation of vanishing layer distance and no tunneling the Hamiltonian can be mapped onto a one layer-system with spin degree of freedom.

In this article we investigate mainly the spin magnetization of a 2DES. From the discussion of the magnetization of a non-interacting system we conclude that interaction is not only a fundamental ingredient for an understanding of the ground state but also of thermodynamic quantities. We outline a recent diagrammatic approach for filling factor one, which takes into account the low-lying spin-wave like collective excitations. [7,8] In Section 4 we discuss other theoretical approaches, in particular the mapping of the system onto a model of a

Heisenberg ferromagnet and compare these theoretical results with experimental data.

2   Elementary Theories for the Spin Magnetization

In the following we discuss as an example of a thermodynamic quantity the spin magnetization in a system of interacting electrons with spin degree of freedom in the lowest orbital Landau level. The Hamiltonian can be written in the Landau gauge $A(r) = B(0, x, 0)$ as

$$H = -\frac{1}{2}\Delta_z(N_\uparrow - N_\downarrow)$$
$$+ \frac{\lambda}{2} \sum_{\substack{p,p',q \\ \sigma,\sigma'}} \tilde{W}(q, p - p') c^\dagger_{p+\frac{q}{2},\sigma} c^\dagger_{p'-\frac{q}{2},\sigma'} c_{p'+\frac{q}{2},\sigma'} c_{p-\frac{q}{2},\sigma} , \qquad (1)$$

where $\lambda = e^2/(4\pi\epsilon\ell_c)$ is the interaction coupling constant, $p = k_y$ is the momentum in y-direction characterizing the degeneracy of each orbital Landau level (the $q = 0$ term is excluded from the sum in order to consider a positive neutralizing background) and $\tilde{W}$ is the interaction matrix element projected onto the lowest Landau level (LLL). Since the coupling constant λ scales with $\sqrt{B}$ the lowest Landau level approximation of (1) is justified in a range of magnetic field strengths, that is fortunately relevant in current experiments. When we compare the energy scales expressed in temperatures at $B = 10$ Tesla we find that the Landau level distance (180 K) > interaction energy scale λ (160 K) $\gg$ Zeeman energy Δ_z (3 K).

The temperature dependent spin magnetization of free electrons at filling factor ν and Zeeman energy Δ_z is given by

$$M(T, \nu) = \frac{Ng\mu_B}{2} \frac{\sinh(\beta\Delta_z/2)}{z + \cosh(\beta\Delta_z/2)} , \qquad (2)$$

where $M_0 \equiv Ng\mu_B/2$ is the maximum magnetization and the fugacity $z = e^{\beta\mu}$ (μ - chemical potential) is related to the filling factor by

$$z = \frac{1}{(2 - \nu)} \left(\sqrt{(1 - \nu)^2 \cosh^2(\beta\Delta_z/2) + \nu(2 - \nu)} - (1 - \nu)\cosh(\beta\Delta_z/2) \right) \qquad (3)$$

[13], see Fig. 1. In the special case $\nu = 1$ it is $z = 1$ and therefore $M(T)/M_0 = \tanh(\beta\Delta_z/4)$, which reflects the itinerant character, i. e. the property not only to change the spin quantum number but also the degenerate quantum number. This leads to a reduced magnetization when compared with the magnetization $\tanh(\beta\Delta_z/2)$ of a non-interacting, localized spin system.

When we compare this with the experimental magnetization curve at, e. g. , $\nu = 1$ originating from the measurements of the Knight shift in a NMR-experiment

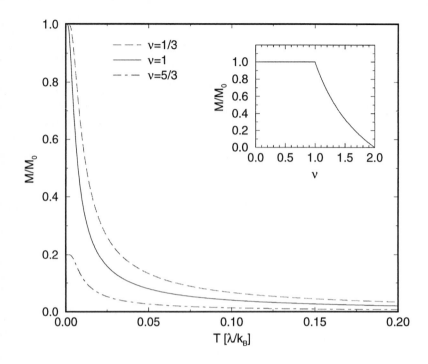

Fig. 1. The temperature dependence of the spin magnetization of non-interacting electrons in the lowest Landau level at filling factors $1/3, 1, 5/3$ when $\Delta_z = 0.016\lambda$. The insert shows the ground state magnetization as a function of ν

[1] we realize a much higher magnetization for any finite T in the experiment, see Fig. 2. Only at zero temperature and non-zero Zeeman term the ground states of the non-interacting and interacting systems coincide because all electrons reside in the spin majority band ($\uparrow$) with spins parallel aligned so that $S^z = N/2$. This happens in the interacting system as the $\Delta_z = 0$ spin multiplett ground state of the quantum Hall ferromagnet with total spin $S = N/2$ caused by quantum-mechanical exchange favors at finite Δ_z the ground state with maximum S^z.

However, for arbitrary ν even the ground states do not agree and therefore the case of $\nu = 1$ is rather an exception than the rule. Obvious examples of the failure of the free particle picture are the strongly correlated ground states at fractional quantum Hall effect filling factors for spin-polarized electrons like $1/3$ and $2/5$. They are truthfully described by the Laughlin wave functions and the Jain wave functions, respectively.

A conceivable step to remedy this unsatisfying result is to treat the interaction in a self-consistent Hartree-Fock (SHF) approximation which leads to temperature dependent energies for the two-spin directions. The energies appear as poles in the SHF-Green's function $\xi_\sigma^{HF} = \epsilon_\sigma^{HF} - \mu$. While for the non-interacting case the bare difference between the spin down and spin up energy is Δ_z, the zero temperature difference $\xi_\downarrow^{HF} - \xi_\uparrow^{HF}$ equals $\lambda a(0) + \Delta_z$, ($\lambda a(0)$ is the exchange energy) and is dominated by the interaction strength. This results in the well-known temperature dependent exchange enhanced spin gap at zero temperature, which reflects at $\nu = 1$ the energy that is necessary to create a neutral quasielectron-quasihole pair far apart (particle-hole gap).

As an illustration, Fig. 2 depicts the magnetization in the SHF approximation at filling factor $\nu = 1$. In contrast to the free particle case, the SHF exceeds the experimental result for the magnetization and we believe that the magnetization from the SHF and of the non-interacting case form a upper and lower bounds for the true behavior. The failure of this theoretical approach becomes evident since it predicts for $\Delta_z = 0$ and temperatures $T \leq T_c = \lambda a(0)/(4k_B)$ a finite spin magnetization. However, this is typical for a mean field theory in dimension two.

The obvious failure of the non-interacting picture as well as of the SHF-theory makes it necessary to go beyond such effectively one-particle theories. Since we are interested in a theory that reflects properly the low-temperature behavior, the case of the filling factor $\nu = 1$ is particularly suited to our needs. There, the ground state is exactly described within the SHF and the low-lying excitations can be derived from a time-dependent Hartree-Fock approximation. Other amenable situations occur at those filling factors, where the ground state is spin-polarized and where the ground state wavefunction and the low-lying excitations are known.

Therefore we concentrate ourselves on the half-filled band case $\nu = 1$. There exists a class of one-spin flip excitations $|k>$, which are exact eigenstates of (1) and which can be labelled by the quantum number k with

$$|k> = \frac{1}{\sqrt{N}} \sum_q e^{-iqk_x \ell_c^2} c_{q,\downarrow}^\dagger c_{q+k_y,\uparrow} |\Psi_{\nu=1}> , \qquad (4)$$

where $|\Psi_{\nu=1}>$ is the spin-polarized $\nu = 1$ ground state. The corresponding eigenenergies are

$$\epsilon_{SW}(k) = \Delta_z + \lambda(a(0) - a(k)) = \Delta_z + 4\pi\rho_s \ell_c^2 k^2 + O(k^4) \qquad (5)$$

which describe for small k collective spin waves with a spin-stiffness $16\pi\rho_s = \lambda a(0)$. In general, Eq. (5) expresses the loss of Zeeman energy Δ_z and exchange energy $\lambda a(0)$ due to one flipped spin of an electron and the attractive interaction $-\lambda a(k)$ between a minority-band electron $\downarrow$ and a majority-band hole $\uparrow$ separated by $\ell_c^2(e_z \times k)$, emphasizing the single-particle character at large k. At $k \to \infty$ the bandwidth of the excitations becomes $\Delta_z + \lambda a(0)$, the zero temperature SHF result (later, we will replace the quantity $a(k)$ by the effective interaction $\tilde{a}(k)$ discussed in connection with screening effects).

If one treats the excitations in (4) as bosonic spin waves, the temperature dependent spin magnetization is governed by thermally excited magnons. The assumption of no interaction between the magnons leads to a theory that is

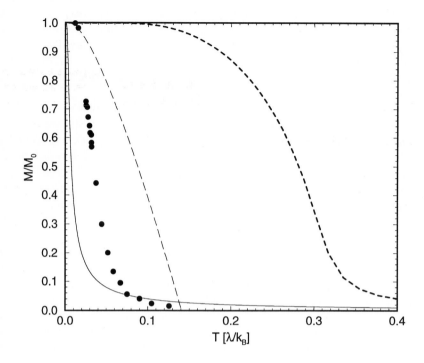

Fig. 2. The temperature dependence of the spin magnetization at $\nu = 1$ in the self-consistent Hartree-Fock approximation (dashed line) and the result assuming free magnons (long-dashed curve). It is $\Delta_z = 0.016\lambda$. For comparison, the results for non-interacting electrons (solid line, see also Fig. 1) and of the experiment [1] (filled circles) are added

formally equivalent to Bloch's theory for the low-temperature magnetization in in a three-dimensional ferromagnet with localized spins, but in our case the dimension is two and Δ_z acts as an external magnetic field. We get

$$M(T) = M_0 \left[1 - \frac{k_B T}{4\pi\rho_s} \ln\left(\frac{1}{(1 - e^{-\beta\Delta_z})} \right) \right] . \tag{6}$$

The linear T-prefactor is the equivalent of the $T^{3/2}$-prefactor in Bloch's law. This rather simple theory leads to a negative magnetization at sufficiently high temperatures, but reflects properly for $k_B T \ll \Delta_z$ the activated behavior of spins to overcome the energy gap Δ_z, i. e. $M(T)/M_0 = 1 - k_B T e^{-\Delta_z/(k_B T)}/(4\pi\rho_s)$, and in the temperatur range $\Delta_z \ll k_B T \ll 4\pi\rho_s$ it is $M(T)/M_0 = 1 - k_B T \ln(k_B T/\Delta_z)/(4\pi\rho_s)$, see Fig. 2.

However, one should keep in mind that a direct comparison of a localized spin system, *e. g.* of a Heisenberg model, with an itinerant electronic system like this one has some conceptual difficulties.

3   A Diagrammatic Approach to the Thermodynamics at $\nu = 1$

Now let us sketch a diagrammatic approach that calculates in a more systematic manner corrections to the electron self-energy and allows the determination of thermodynamic quantities.

Our theory assumes the SHF-propagators as unperturbed propagators, *i. e.* , $\mathcal{G}_\sigma^{HF}(i\nu_n) = 1/(i\hbar\nu_n - \xi_\sigma^{HF})$ (ν_n are the fermionic Matsubara frequencies of the thermodynamic Green's function), and calculates the corrections $\tilde{\Sigma}_\sigma(i\nu_n)$ to the SHF-self-energy $\Sigma_\sigma^{HF} = \xi_\sigma^{HF} - \xi_\sigma^{(0)}$, so that the Dyson equation reads

$$(\mathcal{G}_\sigma(i\nu_n))^{-1} - (\mathcal{G}_\sigma^{HF}(i\nu_n))^{-1} = -\tilde{\Sigma}_\sigma(i\nu_n). \tag{7}$$

The self-energy is derived from the four-point scattering vertex $\Gamma_{\sigma,-\sigma}$, which is approximated by an infinite sum of ladder diagrams taking into account the successive scattering of an electron with spin σ and of a hole with opposite spin $-\sigma$. This leads to a Bethe-Salpeter equation that can be solved analytically due to the independence of the Green's functions of the degenerate quantum number. Closing the $-\sigma$-lines of the vertex yields an explicit expression for the correction of the self-energy for, *e. g.*, the spin direction ↑:

$$\tilde{\Sigma}_\uparrow(i\nu_n) = \lambda^2(\nu_\uparrow^{HF} - \nu_\downarrow^{HF}) \int_0^\infty d(\frac{k^2\ell_c^2}{2})\tilde{a}^2(k) \frac{\{n_B(\tilde{\epsilon}_{SW}(k)) + n_F(\xi_\downarrow^{HF})\}}{(i\hbar\nu_n + \tilde{\epsilon}_{SW}(k) - \xi_\downarrow^{HF})} \tag{8}$$

Eq. (8) provides a nice interpretation of the approximation we employed. Effectively, it describes a coupled electron-spin-wave system when we compare our results with the self-energy of an electron-phonon system calculated in second order of the coupling constant [9]. Thus $\lambda\tilde{a}(\boldsymbol{k})$ plays the role of the electron-spin-wave coupling constant. Therefore, our approximation describes the scattering of an electron with spin ↑ on a bosonic spin-wave with spin -1 into an intermediate electron state of spin ↓ associated with an absorption of one spin-wave. Due to the conservation of the spin the opposite process, the emission of a spin-wave, is not possible for an electron in the majority spin-band ↑. This explains why only one term in Eq. (8) instead of two as in the electron-phonon model occurs where such a constraint does not exist. Reversely, for $\sigma =\downarrow$, no absorption, but only the emission of a spin-wave is allowed. The bosonic distribution function of the one spin-flip excitations $\tilde{\epsilon}_{SW}(k) = \Delta_z + \lambda(\nu_\uparrow^{HF} - \nu_\downarrow^{HF})(\tilde{a}(0) - \tilde{a}(k))$ enters (8), which is similar to (5), but with a temperature dependent band width renormalized by the prefactor $(\nu_\uparrow^{HF} - \nu_\downarrow^{HF})$. Hence, with increasing temperature the probability to flip a spin becomes larger and the magnetization drops faster than without this prefactor.

Within this approximation the calculation of the spin magnetization is easily possible after the spectral function $A_\sigma(E)$ has been extracted from the self-energy expression via the imaginary part of the Green's function:

$$A_\sigma(E) = -2ImG_\sigma^{ret}(E) = -2Im\frac{1}{(E + i\delta - \xi_\sigma^{HF} - \tilde{\Sigma}_\sigma^{ret}(E))} \,. \tag{9}$$

The explicit discussion of (9) reveals that except for the temperature dependent interval $\xi_\uparrow^{HF} \leq E \leq \xi_\downarrow^{HF} - \Delta_z$, where incoherent band contributions occur, the denominator is real. Two quasiparticle poles of the Green's function $E_\uparrow^- \leq \xi_\uparrow^{HF}$ and $E_\uparrow^+ \geq \xi_\downarrow^{HF} - \Delta_z$ can be found as solutions of the equation $E - \xi_\uparrow^{HF} = \tilde{\Sigma}_\uparrow^{ret}(E)$. They show up as quasiparticle poles with weights $z_\uparrow^-$ and $z_\uparrow^+$ in the spin-up spectral function. A more detailed study shows that these two poles exhaust most of the spectral weigth for any temperature: $z_\uparrow^+ + z_\uparrow^- \simeq 1$. In the limit of zero temperature $z_\uparrow^- = 1$ and $E_\uparrow^- = \xi_\uparrow^{HF} = -(\Delta_z + \lambda\tilde{a}(0))/2$, so that the SHF-solution is recovered.

The spin magnetization is given as the ratio of expectation values for the particle number of spin-up and spin-down electrons: $(N_\uparrow - N_\downarrow)/(N_\uparrow + N_\downarrow) = (\nu_\uparrow - \nu_\downarrow)/(\nu_\uparrow + \nu_\downarrow)$, where $\nu_\sigma = N_\sigma/N_\Phi$. Expressing the magnetization by means of the spectral function and using the special particle-hole symmetry at filling $\nu = 1$ it is

$$M(T)/M_0 = \int_{-\infty}^{\infty} \frac{dE}{2\pi}(2n_F(E) - 1)A_\uparrow(E) \,. \tag{10}$$

When analyzing the contribution of the collective excitations to $\tilde{\Sigma}_\uparrow(E)$ from the small k-region in $\epsilon_{SW}(k)$ with a cut-off at $k\ell_c \simeq 1$, we find that the weight equals

$$z_\uparrow^- = [1 + \ell_c^2 \int_0^{l_c^{-1}} dk k n_B(\epsilon_{SW}(k))]^{-1} \,. \tag{11}$$

The location of the low energy pole is then

$$E_\uparrow^- = \xi_\uparrow^{HF} - \lambda\tilde{a}(0)[1 - z_\uparrow^-] \,. \tag{12}$$

Substitution of $A_\uparrow(E) \simeq 2\pi z_\uparrow^- \delta(E - E_\uparrow^-)$ into (10) yields again the low-temperature behavior of the magnetization already known from the naive spin-wave theory (6).

In Fig. 3 the result of the numerical evaluation of the spectral function of our theory is rendered. We choose $\Delta_z = 0.016\lambda$ as a typical value at about 7 Tesla used in the experiments of Barrett et al.[1] It yields a considerable improvement when compared with the self-consistent Hartree-Fock theory and with the naive spin-wave theory.

However, our calculation trying to cover the whole temperature range has some deficiencies. Firstly, the spin-wave-spin-wave interaction is not accounted for as can be seen from the infinite lifetime of the spin-waves at finite temperatures in $\tilde{\epsilon}_{SW}(k)$. From results of exact diagonalizations of a finite number of particles, we realize that an eigenstate consisting of two one-spin-flip spin-waves has a lower energy than the sum of the single spin-wave energies indicating the lowering of the eigenenergy due to interaction and therefore fostering a further decrease of the magnetization.[8] Secondly, screening has to be considered in our approach. Although at zero temperature screening is absent due to the charged excitation gap, for any finite temperature a non-zero temperature dependent screening wavevector k_{sc} exists. We take this into account in a Thomas-Fermi theory, which leads to an effective interaction $\tilde{V}(\boldsymbol{k})$ in the long-wavelength limit, i. e. the Fourier transform of the bare interaction $(2\pi\ell_c)/k$ is replaced by the effective expression $2\pi\ell_c/(k+k_{sc})$. The resulting self-consistent equation for k_{sc} was solved for some temperatures in order to check the influence of the screening on the magnetization. The curve of our theory in Fig. 3 is for $k_{sc} = 0.01\ell_c^{-1}$ and corresponds to a screening wavevector at temperature $T \sim 0.09/k_B$. At least, for low and moderate temperatures the influence of screening is relatively weak, but becomes stronger at higher temperatures. In any case, screening has to be considered at any temperature because otherwise certain self-energy diagrams would diverge logarithmically. Hence, the bare interaction $a(\boldsymbol{k})$ in Eq. (5) has to be replaced by the renormalized function $\tilde{a}(\boldsymbol{k})$.

A different self-consistent diagrammatic method was developed recently by Haussmann.[10] It is particularly suited at high and moderate temperatures, but fails at low temperatures where the magnetization diverges. It is based on a bosonic random phase approximation like theory also starting from the Hamiltonian (1), see Fig. 3.

4   Other Theoretical Approaches and Comparison with Experiment

Another approach to determine the spin magnetization at $\nu = 1$ is based on the observation that the long-wavelength spin-wave excitations of the microscopic model (1) and the low-lying excitations of a ferromagnetic Heisenberg model are similar. However, one should keep in mind, that such a mapping cannot conserve all features of the original electronic model. The Heisenberg model on a two-dimensional lattice with an external magnetic field reads

$$H = -J \sum_{<ij>} \boldsymbol{S}(i)\boldsymbol{S}(j) - g\mu_B B^z \sum_i S^z(i) \qquad (13)$$

with the three-dimensional spins $\boldsymbol{S}(i)$ and the constraint $\boldsymbol{S}^2(i) = S(S+1)$ for each lattice site i and with $S = 1/2$ in our case. The nearest-neighbor exchange constant $J > 0$ of this phenomenological model can be related to the spin stiffness in (5) of the microscopic Hamiltonian (1) due to the relation $J = 4\rho_s$ so that $J/\lambda = \tilde{a}(0)/(4\pi)$. This results in a maximum value of $J = 1/(4\sqrt{2\pi})\lambda$ for the unscreened Coulomb interaction. On one hand, the evaluation of the magnetization can be done by turning to a continuum model, generalizing the orginal $SU(2)$-symmetry represented by two types of Schwinger bosons to a

$SU(N)$-symmetry and evaluating the large N limit.[11] Then, the mean field solution becomes exact for $N \to \infty$. A similar kind of generalization can also be performed, when one starts from a $O(3)$-representation for the spins. The resulting mean field solution for the $SU(N)$-symmetry is equivalent to the approximation of non-interacting magnons in the low-temperature region, see (6) and gives better results than the result of the mean field $O(N)$-theory. A systematic improvement of these theories can be achieved calculating $1/N$-corrections. [12] However, for the $SU(N)$ theory the validity is restricted to not too large temperatures, otherwise the magnetization becomes negative. At moderate and higher temperatures the $O(N)$-theory with $1/N$ corrections is rather appropriate to describe the magnetization. These results can be independently checked by utilizing results of Monte Carlo calculations of the magnetization for the lattice Heisenberg model.[12]

To date, two experimental methods are available to examine the spin magnetization at small filling factors. Firstly, NMR-experiments allow the measurement of the electron spin magnetization because the electrons generate an additional magnetic field at the location of the ^{71}Ga-nuclei, which is proportional to the measured Knight shift of the resonance line. Assuming saturated ferromagnetism at $\nu = 1$ the relative magnetization can be derived from the strength of the shift. This technique was put forth by Barrett and collaborators and provided data for the temperature dependent spin magnetization at and about $\nu = 1$ and 1/3 as well as data for the spin-lattice relaxation rate $1/T_1$.[1,6] Secondly, Goldberg et al. derived directly $M(T)$ from magneto-optical absorption experiments. [2] The measured absorption for optical interband transitions into the majority and minority spin states of the LLL, respectively, is proportional to the calculated optical matrix elements for those transitions and the number of non-occupied states in the final state in the LLL with spin σ. Qualitatively, the results of these two independent methods agree. Therefore, we restrict the comparison of the various theoretical results shown in Fig. 3 to Barrett's NMR-data.

Before comparing the theoretical results with experiment some words of caution are necessary as the two models used to describe the quantum Hall ferromagnet at $\nu = 1$ are originally quite different. From the microscopic point of view the Hamiltonian (1) is much better justified than the Heisenberg model. In general, the latter does not account for the charge degree of freedom and therefore the itinerant character is neglected. This has consequences: the high temperature expansion of the Heisenberg model yields as leading term the same result as that for free localized spins, namely $M(T)/M_0 \sim (\beta\Delta_z)/2$, while $(\beta\Delta_z)/4$ is the correct result at $\nu = 1$, see Eq. (2). On the other hand, such a high-temperature expansion should not be overestimated as mixing of the Landau levels is excluded. A disadvantage of the continuum quantum field theory is its inability to describe experiments where the charge is important, e. g. in tunneling experiments in a bilayer system where the electrons can tunnel through a barrier between the two parallel 2DES which are kept at the same filling factor. Applying a voltage V between the layers *without* changing ν in each layer allows to measure the tunneling current I. Theoretically, this current-voltage characteristic can be expressed as the convolution of the single layer spectral functions whose calculation was done above. Therefore we can predict the temperature dependent $I - V$ characteristic at $\nu = 1$. Such an experiment offers another opportunity to check the quality of the calculated spectral function. [8]

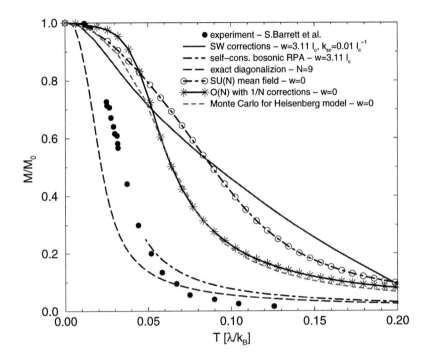

Fig. 3. The temperature dependence of the spin magnetization in various theoretical approximations compared with data from [1] rendered as filled circles, where the Zeeman energy is $\Delta_z = 0.016\lambda$ and the quantum well width is $w = 3.11\,l_c$. The solid line shows our theory based on spin-wave corrections to the electronic self-energy with a static screening wavevector $k_{sc} = 0.01\,l_c^{-1}$, the dot-dashed curve is the self-consistent random phase approximation valid at moderate and higher temperatures [10] and the long-dashed curve depicts the result of an exact diagonalization for nine particles with a width similar to that in the experiment.[8] The next three curves are results for the Heisenberg model without finite thickness corrections which would lead to a further decrease of the magnetization: the dot-dashed curve with circles is the mean field $SU(N)$-theory, the solid line with stars is the result of a $O(N)$-theory with $1/N$ corrections and the dashed line is the result of the Monte Carlo calculation [12]

Comparison with experiment requests the consideration of the finite width of the quantum well perpendicular to the plane so far neglected in the theories. The single particle wave functions exhibit actually also an extension in the direction perpendicular to the layer, $e.$ $g.$ a width of about 30 $nm \simeq 3\ell_c$ in Barrett's experiment, and the interaction has to be corrected by a form factor. Assuming an infinite heigth of the quantum well and an ideal Coulomb interaction the function $a(\boldsymbol{k})$ becomes smaller and the effective spin stiffness is only one half of the ideal value $\lambda/(16\sqrt{2\pi})$.

Fig. 3 shows the results of the theories discussed above and compares them with the NMR-data. The numerical diagonalization data are, except for finite size corrections, exact and quite close to the experiment.[8] For low temperatures the experimental data decrease too slowly and for large temperatures too quickly since they fall even below the magnetization of free electrons. Our theory developed in section 3 is an improvement over the SHF-theory and the naive theory of free magnons and reflects the correct low-temperature and high-temperature behavior, but is much too high at moderate temperatures where the magnetization drops almost linearly. The results of the Heisenberg model show astonishingly good agreement with experiment. All the theories neglect disorder and Landau level mixing. However, there are some experimental hints that disorder changes the clean-limit behavior and accounts for deviations mentioned above, whereas the influence of higher Landau levels seems to be small.[10] In all theories evaluated so far skyrmions are not taken into account. Although they are the energetically lowest charged excitations, their influence should be small, since the one-flip excitations dominate energetically at small and moderate temperatures, where experiments are performed.

5   Summary

In this paper we studied the thermodynamics of a quantum Hall ferromagnet calculating the temperature dependent spin magnetization at $\nu = 1$. At least for this filling factor, we could show that an effectively one-particle treatment is not sufficient and that a many-particle approach is necessary that accounts for the collective spin-wave excitations. However, such a situation does not always occur as recent experiments [6]and theoretical studies [13] of the quantum Hall ferromagnet at $\nu = 1/3$ indicate. There, the ratio of the particle-hole gap to the Zeeman gap is much smaller than that at $\nu = 1$ so that the low-temperature physics can be described by the $\nu = 1$ free particle formula with a renormalized Zeeman gap. Besides the approximate calculation of the spectral function, in principle, also the determination of two-particle quantities like the spin susceptibility, which enters the spin-lattice relaxation rate as a measurable quantity, is possible. In concluding, we note that the 2DES in a strong magnetic field is particularly appropriate to test theoretical ideas with importance for itinerant ferromagnets due to its close relationship to experiment.

Acknowledgements

I would like to thank Allan MacDonald for introducing me to this field and giving me the opportunity to share insights with him on this fascinating subject. I wish to thank J.J.Palacios for providing the data of the exact diagonalization. Valuable discussions with S.Barrett, S.M.Girvin, B.Goldberg and R.Haussmann are gratefully acknowledged.

References

1. Barrett, S.E., Dabbagh, G. et al. (1995) Optically pumped NMR evidence for finite-size skyrmions in $GaAs$ quantum wells near Landau level filling factor $\nu = 1$. Phys. Rev. Lett. **74**, 5112–5115
2. Manfra, M.J., Aifer, E.H. et al. (1996) Temperature dependence of the spin polarization of a quantum Hall ferromagnet. Phys. Rev. B **54**, 17327–17330
3. Wiegers, S.A.J., Specht, M. et al. (1997) Magnetization and Energy Gaps of a High-Mobility 2D Electron Gas in the Quantum Limit. Phys. Rev. Lett. **79**, 3238–3241
4. Meinel, I., Hengstmann, T. et al. (1999) Magnetization of the Fractional Quantum Hall States. Phys. Rev. Lett. **82**, 819–822
5. Bayot, V., Grivei, E. et al. (1996) Giant low temperature heat capacity of $GaAs$ quantum wells near Landau level filling $\nu = 1$. Phys. Rev. Lett. **76**, 4584–4587
6. Khandelwal, P., Kuzma, N.N. et al. (1998) Optically pumped nuclear magnetic resonance of the electron spin-polarization in $GaAs$ quantum wells near Landau level $\nu = 1/3$. Phys. Rev. Lett. **81**, 673–676
7. Kasner, M., MacDonald, A.H. (1996) Thermodynamics of Quantum Hall Ferromagnets. Phys. Rev. Lett. **76**, 3206–3209
8. Kasner, M., Palacios, J.J., MacDonald, A.H. (1998) Itinerant Electron Ferromagnetism in the Quantum Hall Regime. cond-mat/9808186
9. Mahan, G.D. (1990) Many-Particle Physics, p. 176. Plenum Press, New York
10. Haussmann, R. (1997) Magnetization of Quantum Hall Systems. Phys. Rev. B **56**, 9684–9691
11. Read, N. and Sachdev, S. (1995) Continuum Quantum Ferromagnets at Finite Temperature and the Quantum Hall Effect. Phys. Rev. Lett **75**, 3509–3512
12. Timm, C., Girvin, S.M. et al. (1998) $1/N$-expansion for two-dimensional quantum ferromagnets. Phys. Rev. B **58**, 1464–1484
13. MacDonald, A.H., Palacios, J.J. (1998) Magnons and Skyrmions in Fractional Hall Ferromagnets. Phys. Rev. B **58**, 10171–10174

Printing: Druckhaus Beltz, Hemsbach
Binding: Buchbinderei Schäffer, Grünstadt